Don't Throw IT

~

Get To Know IT!

A Computer Guide
for the
Technically Challenged

J. Wiltz Cutrer, Jr.

Don't Throw IT ~ Get To Know IT

Please direct requests for permission to make copies of any part of this work or to request the author speak at your event to:

J. Wiltz Cutrer, Jr.
206 Kitty Hawk Circle
Brandon, MS 39047

601-706-9589

KnowIt@TechKnolutions.com

Visit us on the web:
www.TechKnolutions.com

Cover Clipart by Ron Leishman - http://clipartof.com/1044922
Back Cover Clipart by Ron Leishman - http://clipartof.com/5761
Inside Cover Clipart by Ron Leishman - http://clipartof.com/438538

ISBN 978-0-9885928-0-3

Library of Congress Control Number: 2012953396

First Edition | 2013
Second Edition | 2017 (*minor updates*)

Printed in the United States of America

Published by TechKnolutions, LLC, Brandon, MS

This Book is Dedicated
To
My Wonderful Son,
Wiltz III,
and
My Magnificent Daughter,
Izabella.

There is no greater joy in a father's life
than to see his children grow and shine.

Every day is a blessing to me as
I get to see your smiles,
hear your laughter,
and feel your hugs.

Truly, my greatest joy
is the sound of hearing
you call me
"Daddy"!

∞ ∞ ∞ ∞ ∞

"Home computers are being
called upon to perform
many new functions,
including the
consumption of homework
formerly eaten by the dog."
– Doug Larson

∞ ∞ ∞ ∞ ∞

Acknowledgements

The author would like to thank so many people that were invaluable to making this book a reality.

First and foremost, I must thank my wonderful wife, Kristen. Throughout the years, the tears, and the crazy looks she has always stood by my side *(or right behind me when I needed a swift kick...)* I Love You, Babe!!!

When it came to giving this book a title I found out just how hard things could be. Even though I had all kinds of crazy ideas, my brother Knight, Mike Brechtel, helped give me the direction needed to get the book title under control and on the cover.

Editing…oh did my writing need a little correcting. Thankfully, I had plenty of help in this regard. A very special thanks to these dear friends: Candice Harris, Janet Martin, and Bevin Tyler.

I had many great conversations "*brainstorming*" this book idea and I must thank those who helped me organize this cluttered space I call my mind: Tony Cox, Charlie Case, John Helms, and Adelle Martin…please know that I greatly appreciated your council! ☺

So many others played such a role in helping me with content ideas. I cannot begin to remember them all. Their inspiration may have come from posts they made on Facebook, comments through conversations, lessons during classes, or from times spent fixing their computer. The people upon whom I call my "*friends*" are vast and wonderful. No matter the distance, know you are special to me, as well as included within this special acknowledgement.

Table of Contents

Prologue

Does the world REALLY need another computer book? As I started writing, I asked myself this very question. Obviously, I concluded that YES we do! Throughout my career, I noticed an aspect of technology that seems to be lost in many other books discussing technology. Many make assumptions about people's exposure to knowledge. They treat the reader like the kid in class that does not ask the question and instead suffers in silence while getting more confused. They then move on with the lesson presuming everyone understood. I wanted to tackle technology from a different angle altogether. After all, we REALLY require very little from our technology to keep us satisfied and productive, and we will address each of these needs:

- We want it to turn on when we need it
- We want it to do what we ask of it in a timely manner
- We want it to give us information when we ask for it
- We want it to turn off without errors

One area where I have seen the user neglected throughout my career has been in their treatment by other computer nerds. I see users told, "*Because I said so*" when we introduce a new policy or procedure. This approach is frustrating to my kids, so why would I force it on my co-workers? I believe it is time we take an educational approach when dealing with users. We should not simply dictate policies but instead facilitate understanding. An educated computer user is going to be a much stronger ally. If we can develop understanding between us, we can actually make policies and procedures that matter!

My principal objective is to discuss technology in down-to-earth terms so everyone can be included and not feel lost after you turn on the power switch. Whether you are an old pro or a relative newcomer, there is something in here for you. I will work hard to answer those questions you did not get a chance to ask. Beyond just telling you what you SHOULD be doing, I will explain WHY and HOW it may help you.

So much we do today is tied to this digital giant we call the computer, and to its first cousin, the Internet. Not even buying a cheeseburger is possible without a computer involved. Some may argue technology today is a complicated and cumbersome mechanism. Coming from an area impacted by that little thunderstorm in 2005 known as Hurricane Katrina, I noticed very fast that life without a computer, and electricity running that little box, was difficult. Those debit cards we count on do not work so well without power and connectivity to run them. Long gone is the familiar clickity-clack of the sliding, manual credit card thingy (*how's that for technical terminology*?) Oh, and do not even get me started on the lost art of "*making change*" when handed cash…without the register to do the math! For that matter…who is still carrying cash today? You know…that green paper that folds in your wallet.

So…good or bad, it is what we have to work with today. So why did I decide to develop this book? It has long been my goal to remove the veil of mystery from these technological foes. I want to help everyone understand the technical tools literally sitting at your fingertips. Perhaps people will even learn to "*make peace*" with their computer. There are many advantages with technology, and I would love to see you make the best of them…and have fun along the way.

Through the upcoming chapters, I will show by example what I have learned throughout my career as the "*IT Guy*". There have been plenty of laughs, a few tears, and a whole lot of things you just cannot make up. I will pour these experiences into these pages and bring you along my journey to "*nerd-vana*", liberation from the ill will of your technology demons!

I have based this book on examples from the Microsoft® Windows® operating system. I know there are some of you reading this which use or Apple® or Linux. While the specifics may not be the same, the principles surely will still apply. At the end of the day, it is all about operating our systems efficiently. It is just like driving a car! While the buttons and knobs may be in different places, and the gas cap may be on different sides, they still really drive the same!

Throughout the chapters, we will discover what it really means to have computer security. We will discuss, in easy-to-relate language, what us nerds mean when we start talking about viruses and other cooties. We will cover home networking, including troubleshooting those annoying times when you cannot get online. Moving on, we will talk about ways to protect you and your computer while surfing around in the World Wide Web. Additionally, if that is simply not enough for you, we will then chat about some frequently asked questions to find the underlying cause of some of the more common issues in this digital age. To close, we will take some techie terms and applied a common-sense definition to them to tear down the digital barrier.

As I was writing, I felt a little like the magician who reveals the trick behind the "*magic*". In no way do I intend for this book to bypass your friendly neighborhood geek shop. There are times when you truly need professional help. (*Some more than others!)* Nevertheless, what I do hope to do is keep you from running out and dropping your hard earned cash (*remember, it is that green stuff that hides in your wallet*) every time a technical issue arises. By gaining a better understanding of the tools at your fingertips, I believe we can move many things away from the *headache* column. Obviously, you must think so as well or you would not have read this far through so many lame attempts at humor.

DISCLAIMER:

While I have made every effort to insert and apply FACTS,
it would not truly be possible without the insertion of opinion
as well. Some readers may still disagree with solutions discussed.
Guess what? That is AWESOME! We are all entitled to such.
My opinions and solutions are the result of MY experiences,
just as yours are…well…yours!
Furthermore, I have received NO compensation from
companies for endorsement of their products or technologies.
I am simply sharing my experiences in the hope I can help.

Computer inSecurity

*"Companies spend millions of dollars
on firewalls, encryption and
secure access devices, and its money
wasted, because none of these measures
address the weakest link in the security chain."*

– Kevin Mitnick

*(You are addressing the weakest link by reading this book...
EDUCATION about technology will truly help you become safer!)*

Does anyone REALLY need an Anti-Virus program? What does Anti-Virus REALLY do? Why is it important? Can you live without it? Is Malware different from Viruses? Are hackers REALLY a threat to little ole me? Can I become infected, even if I do not visit "*naughty*" sites or open strange emails?

Yes; Lots; Because I Said So; Yes; Yes; You Betcha; YES!

Before we can dive into Anti-Virus, we first need to define a few things. Just what IS a virus? We see that term thrown around quite a bit. It seems to be used like a catchall; an excuse when we have no other explanation. Makes me think of the way we generalize when one of our kids isn't feeling well…." *Oh, it's a cold…a touch of the flu…something they ate*". However, it is MUCH more than that, folks.

Before we go too far, there is another question I hear many times I feel we need to address early. Why do people MAKE viruses? There are many opinions out there, so let's discuss a few of them:

- **It is the Anti-Malware companies making them to stay in business:** This is a common "*conspiracy theory*" I blow right by, but still wanted to address. I do not see where this is highly likely. It would not be in any company's best interest. If this were remotely true, I believe it would have been publically revealed LONG ago. This would cause irreparable damage to a company, likely putting them out of business.

- **"Look What I Can Do":** Now, this theory I do believe. We all likely remember, or were, the kid in class that wanted to impress someone they had their eye on. How did we try to catch their eye? By doing something stupid. As a guy I can relate to this rather well. I think quite a few malicious programs are written by someone trying to show another person what they are capable of doing. In fact, the first personal computer virus, named "*Elk Cloner*" was released in 1981 as a practical joke.

- **Just Plain Mean:** Plain and simple, some people are just mean and destructive. Add that to a brilliant mind and you end up with creativity that can literally bring the connected world to a crawl in a matter of moments.

- **Activism through Hacktivism:** Watch the news today and you will see cases where malicious software or web site defacements are occurring to further a cause. We even see instances where countries are caught using these same tactics against others. There is a whole movement of people that use the Internet to further their point of view.

- **Oops:** There is always what I call the "*Steve Urkel*" phenomenon: "*Did I do that*?" Mistakes happen. A mistake in writing a program can sometimes backfire with unintended consequences. (*For those not understanding the Steve Urkel reference, please Google®* "Family Matters*", a show popular during the 90's*).

NOTE: *For those wishing to be technical, the first recognized "virus" was actually written and released 10 years earlier as an experiment. Its name was "Creeper".*

So, What Exactly IS a Virus?

A "*virus*", as we so commonly refer to it, is actually just one component of a much larger collection of unpleasantness: Malware! Malware or **Mal**icious Soft**ware** is a collection of nasty little cooties that can get into your computer and cause all sorts of bad things to happen. Just like the viruses we fight in a medical sense, each form of Malware has a unique set of symptoms. They can result in the slowing down of your computer, loss of data, or even the using of your system to spread the Malware to others.

Virus. Princeton University online dictionary defines a virus as: *a segment of self-replicating code planted illegally in a computer program, often to damage or shut down a system or network.* What does that mean IN ENGLISH? It states the virus is capable of spreading itself with the intent to cause damage to any systems it infects. Sounds a lot like the flu, doesn't it? Like the flu, it can change over time. As soon as a way is found to defeat one strain, we see a new variant come on the scene. Through the years there have been thousands, if not millions, of viruses attacking your computer…each one potentially worse than the last.

Trojan Horse. Like the tale of the Greek horse used to invade the city of Troy, a Trojan Horse can appear harmless, or even helpful, yet hide a deadly secret inside. This infection mechanism attaches itself to a "*seemingly*" harmless file. It could even attach to a PDF or picture. When you open this "*seemingly*" harmless file, you get to see its content…the picture for example. NOW comes the malicious part. In the background, the malevolent action also executes and you unknowingly infect your computer. Since you opened the program, it runs with the same privileges you have on the computer. That means if you are running as a local administrator (*as most users are*) it runs with full privilege to modify any computer settings!

Worms. No, I am not talking about the creepy crawlers in your tomato garden. These devious little critters sneak through your computer and move through the wires to infect others you are connected to across your local network and the Internet. Much like viruses, you do not have to take any action to become infected. Just being there and connected can be enough. Moreover, once they get into your computer they start looking for ways to spread. They do not even require your help. They can do it all by themselves. That means the activities that cause the initial infection do not have to even BEGIN with you. Someone else in your home or office could bring in the infection and it will then rapidly spread to everyone! We always heard sharing was good…but not in this case!

So, what do worms DO? It seems typically their "*payload*", or result, is simply to disrupt network operations. They slow things down and cause a general disruption to your day, and your efficiency. They are like that annoying kid in class who really doesn't contribute much but makes you late because they wasted the teachers time.

Spyware. Wow! Spying? Can your computer REALLY be spying on you? Well, do not break out the tinfoil hats, but the answer is yes! Little things can run in the background tracking your activities and reporting them back to others. This could be to send you targeted advertisements while you surf the web. If the spyware realizes you spend a lot of time on sites looking for shoes you will run the risk of getting athlete's foot from all the shoe-related ads appearing while simply surfing the web! This may seem safe enough, or even helpful.

However, sometimes the software can take it further. Some spyware can do something called "*key logging*". In this instance, your key strokes are recorded for delivery to another party. This can reveal things such as passwords or personal messages. Let that sink in for a moment! Wow! Could someone really get a copy of everything I type on my computer? Yes, they can.

Spyware can also redirect your Internet browsing. This helps the malicious person direct you to the information THEY want you to see. This may be to sell their items. It could be to redirect you to infect your computer with more malicious software. As this continues, you can literally see your computer slow to a crawl.

Spyware is also capable of taking control of your computer. Wait? What? Take CONTROL of your computer? Yes! There are ways that someone can harvest your resources for their own gain. They could then use your computer to send out their information and attempt to infect others. By infecting your computer, you could be added to a remote-controlled group of computers used for criminal activities. This is referred to as a "*Bot Net*", referring to the robotic control of your computer, along with others, across the Internet.

Scareware. OK, I am NOT making this up. Scareware is designed to do just what it sounds like: scare you into taking an action that is not truly necessary. It is, in essence, unethical marketing. It attempts to shock or threaten you into action. It is the "*bully*" of the Malware family.

A common example of Scareware would be a pop-up on your computer indicating you are severely infected with hundreds of malicious items. Moreover, for the low, low price of $29.95 this "*lovely*" program will make all your troubles go away. These can look rather convincing, even imitating some popular security programs. Once you fall for the fraud, the real damage will begin. Besides the fact you may have just given your credit card information to a criminal organization, you may have also just opened up your computer to all sorts of nastiness.

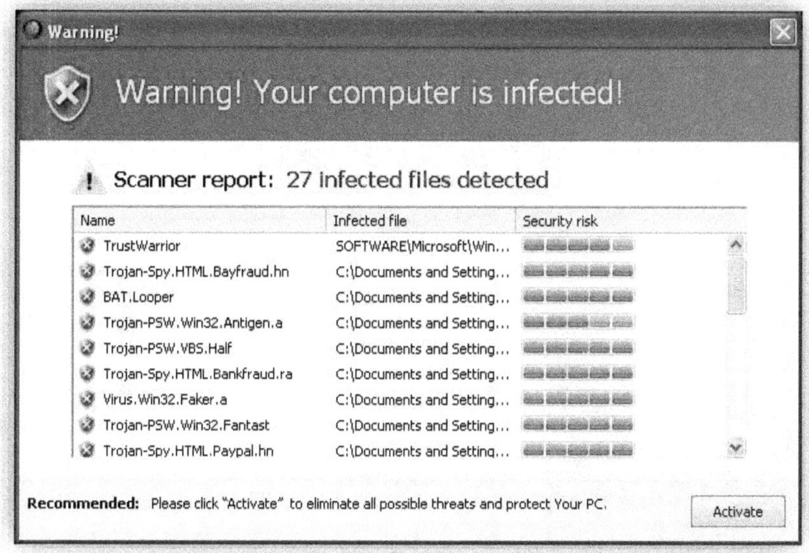

Sample Image of a Scareware Pop-Up

Ransomware. This is a program…a program with a very particular set of skills. Skills it has acquired through years of evolution. Skills that make it a nightmare for computers like yours.

Please tell me at least ONE of you read that in the Liam Neeson voice from "*Taken*"? Much like his daughter in the movie, this is where your computer is held for ransom, although not removed from your physical possession. Your data is encrypted and you are demanded to pay the ransom before you can access your data again. In my mind, this is closely related to Scareware, but with a more sinister, destructive consequence. By encrypting, or locking up, your data your computer becomes useless until you can either remove the infection or pay the ransom. This may not seem a big deal if it just locks up your web surfing records…but imagine if this happened the day before payroll at your business! What if your customer records were unavailable? What if you couldn't get to grandma's secret chocolate chip cookie recipe? I think you get the picture!

What are the dangers of simply paying the ransom? The most obvious issue would be turning over your credit card information to a criminal organization. Identity theft would also be in there as a problem. Beyond that, the likelihood of the paying of the ransom ACTUALLY removing the malicious software is extremely low.

Sample of Ransomware Requiring Payment to Go Away

Wondering what "encrypted" means? Here is how I define it:

***Encrypted** – When data is scrambled into unrecognizable "garble" for the purposes of hiding the information. It requires a "key" to descramble and return to readable form.*

Now we know we are fighting much more than simply viruses. The more accurate term for programs that fight these threats would be Anti-Malware. Ever wonder HOW Anti-Malware programs work? If so, you are in for a treat. If not, you may want to skip ahead!

Anti-Malware programs protect you against the constant stream of malicious software bombarding your systems. These are attacking your web browsing, the applications installed on your computer, and even your computers operating system. Contrary to common belief, vulnerabilities extend far beyond just your operating system. ANYTHING running, or even simply installed, on your computer is a potential target. Anti-Malware programs are there to do their best to prevent these attacks from being successful. There are many ways they accomplish their task:

- Real-Time / On-Access Scanning
- Full System Scanning
- Heuristics

Real-Time / On-Access Scanning. This is the "*active*" portion of Anti-Malware programs. As the program is constantly running in the background when your computer is on, it is in the perfect position to check files and folders as you use them. Think of it as an ever-present watchdog sniffing around. When you go to open a text document or other program on your computer, the Anti-Malware program first checks things out to see if all is well, in its opinion. If a security problem is noticed your security program will let you know. It is important to become familiar with the warning messages your chosen program provides so that you are not tricked by a malicious program or hoax.

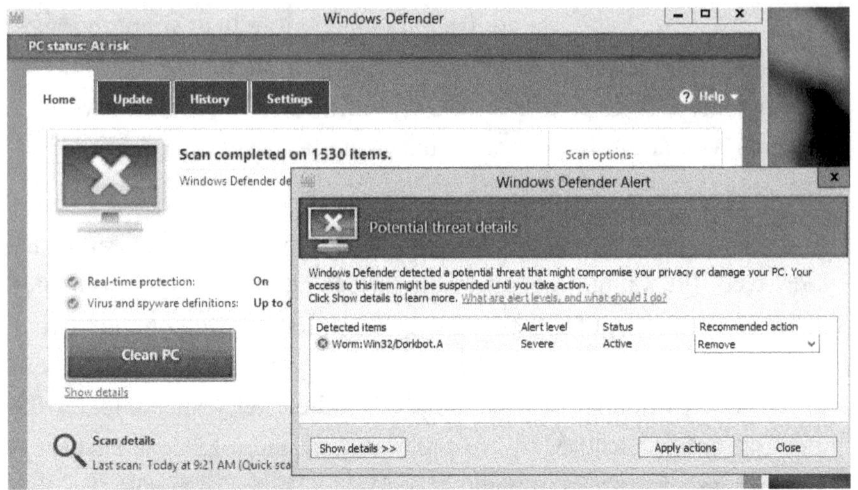

Sample of Detailed Alert from Microsoft® Windows Defender

Full System Scanning. The use of real-time scanning makes the full system scan less urgent to employ. Since the Real-Time scan actively scans items as you open them, it is your primary line of defense. Full System Scan looks behind the scenes and in the corners for rarely touched items. As a result, many typically will only perform this scan on an as-needed basis. I would still recommend this to be set to run at least weekly, as is the default setting for most commercial products. The full system scan will check each file and folder inside your computer and evaluate whether or not it is considered a threat.

∞ ∞ ∞ ∞ ∞

"The Internet? That thing still around?"

– Homer Simpson

∞ ∞ ∞ ∞ ∞

Heuristics. Now there is a 75¢ word. Heuristics is the technology utilized to identify malicious items that may not yet be recognized by the programs definition files. What heuristic scanning does is identify patterns and processes that exhibit malicious-like behavior. Think of this scan as the security guard noticing suspicious behavior and coming over to check things out. They may not be a KNOWN criminal, but seeing them crawling out your window in the middle of the night with a television in their hands sure fits the profile!

One common complaint regarding Anti-Malware software methods is something known as the FALSE POSITIVE. This is where your security software incorrectly identifies legitimate actions or software as malicious. We have to keep in mind that there is an almost unimaginable amount of data throughout the world. No security software is immune from the issue, though some do seem to be less inclined to make these mistakes.

The opposite scanning results can also occur. A FALSE NEGATIVE is when a malicious program manages to slip by your security software undetected, missed by the various scanners. This is most commonly illustrated by spam email slipping through. Remember, no security software is a cure-all. Just like a home or car alarm, you cannot 100% prevent issues. Anti-Malware software is simply a tool to provide information and assistance to your most powerful weapon against malicious software: You!

Preventing Malicious Infections

What do we do to prevent infections? Like so many others year after year, we march out and get our annual flu shot. Yes, I know… not everyone gets a flu shot…and guess what? Not everyone runs Anti-Malware software, though I cannot for the life of me understand why someone would NOT want Anti-Malware software on his or her computer! While there are several good arguments on both sides of the vaccine debate, the Anti-Malware debate is much clearer and less controversial. We could safely say Anti-Malware software is more a requirement than a recommendation.

I will leave the flu shot debate to other books, but for sake of our conversation, let us presume you DO have Anti-Virus. What kind? Well, that is like asking what kind of car you drive. Our answers may vary, but in the end, we essentially hope for the same result, right? Just like our cars, some software is more reliable than others. Some can even cause more trouble than they are worth. How much car is TOO much car? We can apply these same questions to your Anti-Malware.

When I am evaluating security software, I look for a few key components. I have a "*shopping list*" of sorts. There are features I want and some I do not. Sometimes the more bells and whistles the bigger the headache in the end, so I look for a balance.

- How will it affect the performance of my computer?
- What is the reputation of the manufacturer?
- How frequently does it update its definitions?
- How effective is it at catching malicious software?
- What additional features, or protections, does it provide?

How will it affect the performance of my computer? Many times, I hear complaints from people that their security software slows their computer to a crawl. Is this true? Unfortunately, it is. Security programs, and some of the bloated functions that come packaged with them, can cause a significant slowdown in your system. That once blazing, top-of-the-line computer performs suddenly as if a 45 record is being played at 33 speed. Do not worry, younger readers, if you have NO idea what a record is; just ask the nearest person with a little gray hair and they can point you in the right direction.

What is the reputation of the manufacturer? Let us bring back that car analogy. We all know who our common car manufacturers are. Whenever a new company comes on the scene, they can have issues with brand recognition until they prove themselves. Computer software is the same way. You have your industry giants and you have your startups. It is MY opinion that the number of employees on the payroll does not necessarily equate to the quality of their product. Reputation is what we, the people, think of it. This is likely one of the biggest advantages of the Internet age! There are plenty of reviews. Some good…some bad…and somewhere in the middle, the truth. Your best indication of true "*performance*" is what the people using the program are saying.

∞ ∞ ∞ ∞ ∞

"Before you become too entranced with gorgeous gadgets and mesmerizing video displays, let me remind you that information is not knowledge, knowledge is not wisdom, and wisdom is not foresight. Each grows out of the other, and we need them all."

– Arthur C. Clarke

∞ ∞ ∞ ∞ ∞

How frequently does it update its definitions? While there is a little bit of "*best guess*" to security software, at its core are the definitions. Definitions are the set of instructions that are the heart of the security software. They tell the program WHAT to look for in order to identify security threats. While many say ANY Anti-Malware protection is better than none at all, I would suggest that if your definitions in your program are out of date, you in essence ARE running nothing at all, as its effectiveness is significantly hindered. The company develops definition files to address the malicious software they discover across the world. When they identify a new threat, a team of programmers will set to work writing a way to detect and defeat the threat. They deliver these updated actions to your software via definition updates. As hundreds of threats are identified daily, too much of a lapse in updating can cause you quickly to fall behind the threat curve. The manufacturer's responsiveness is vital to staying current. You should research their reputation by reading opinions posted by others. In addition, most manufacturers of security software will openly boast of their methods and frequency of updating.

How effective is it at catching malicious software? Now we are getting down to the details. Does it do what it says it will? Is it effective? Let me first answer this question with a blanket statement: There is **NO** silver bullet when it comes to security software. There is not a program that I or anyone else in the world could sell you that will be 100% effective against current and emerging threats. Moreover, anyone who tells you otherwise is also willing to sell you oceanfront property on the beaches of Tennessee. There is simply NO way to predict what the next threat will be to your computer. Therefore, security software will rely on "*typical*" patterns to predict and prevent new threats, but it cannot do it with 100% certainty. So again, we go to the opinions of the masses, along with some studies conducted regularly on these things. (*Oh...and keep reading...I may have some opinions for you as well*!)

As a way to research a program's effectiveness, also referred to as "*Detection Rate*", you can check out sites online such as AV-Comparatives (**www.AV-COMPARATIVES.org**), which conducts ongoing independent studies comparing current security software. As you will note as you look through multiple reports, effectiveness seems to ebb and flow like the tides. This is due to the ever-changing nature of the computer environment. So do not base your evaluation on one chart, or one month. To see the WHOLE picture takes a bit more of a look under the hood. For example, the program with the highest catch rate this month may also have had a very high false-positive rating last month, indicating that in its aggressive urge to catch everything it makes many mistakes. These mistakes will cost YOU time responding to false positives on your computer.

What additional features, or protections, does it provide? This is where things can get complicated. Extra firewalls, web add-ons, etc. are common extras found in consumer security software. All the bells and whistles may or may not be worth it. Much like cars, it is not a "*one size fits all*" world when it comes to security software. While some will take the "*bare necessities*" approach to be lean and clean, others are more like "*everything AND the kitchen sink*". I have always found the more complicated, high-end "*features*" are very rarely used and amount to nothing more than "*Bloatware*", which ends up serving no purpose other than to clutter things up. Think of all the stuff that may be sitting in your garage right now, and getting in your way, when all you really want is to get to the lawn mower! Yes… that's Bloatware. Therefore, if your security software is full of OOOH and AHHH features you will never really use, you may see more of those performance hits we discussed earlier.

Balance. Hard to achieve? Not really. You just have to know what you are looking for. Do I have opinions? You betcha! I have spent a lot of time playing with, installing, removing, and generally messing around with quite a few different security software programs and I have come up with some favorites. Therefore, at the risk of insulting some of the programs on your computer, I will tell you what I PERSONALLY use to protect my family and friends.

WARNING: You should ONLY have ONE core security program on your computer! While some programs can coexist, some cannot. For example, Norton® and McAfee® together would NOT double your protection, but probably result in crashing your computer, or at the minimum render it virtually a sloth! If, after you install a program, you notice a slowdown, uninstall the program and re-evaluate. Many times a quick online search can provide insight into if the program you installed has known conflicts with existing software.

† † † † †

RANDOM TECH TIP:

Ever wonder why your email gets blocked?
Email filters are automated systems that define messages based on dictionary words and patterns. Typically, they do not know how to distinguish "content". For example, lengthy disclaimers touting no DISCRIMINATION based on color, SEX, or RELIGION would typically cause those capitalized words to be singled out and assign a spam value to them. When the value gets too high, the message is "flagged" as possible spam. This forces it to the head of the pack of thousands for a "human" review. Bulk emails, such as from companies like Constant Contact gain even higher scrutiny based on the fact they "bulk" mail, like advertisements in your regular mailbox.

PICNIC (Problem In Chair, Not In Computer)

Beyond Anti-Malware, there are many other things we need to consider regarding the security and safety of your computer against the threats that exist. Simply slapping on a couple programs is not going to do much. It would be like locking the doors on your car while leaving your windows down. Sometimes it may not be a virus or some hacker at all. Sometimes the problem stares back at you in the mirror!

Oh no! Did I just say YOU might be your own problem? Yes, I did. It happens all too often. We overhear, see an ad, or watch a commercial about the latest whiz-bang, golly-gee, awesome application and we cannot wait to hop on our computer and install it! The problem is that what you are installing IS the problem. Now I am not saying all programs promoted in this way are bad. Far from it. What I AM saying is everything you read may not really be true! Plenty of people are more than willing to take advantage of your frustrations for their own financial gain.

Malware Primary Protection: My primary (*core*) program of choice is Microsoft® Windows Defender, included FREE in current Windows offerings. I have been impressed with its balance of performance and precision when dealing with the assorted cooties that may try to slip into my computer. Microsoft® Windows Defender updates regularly, integrates well with Windows®, causes minimal performance hits, has a very good "*catch*" rate when seeking out newly emerging threats, and is simple and straightforward. Another nice benefit of this program, besides its effectiveness? It is FREE! Yes…free, as in it will not cost you a dime. Moreover, if you are a businessperson, Microsoft® will allow it FREE for any small businesses with up to 10 PCs. (*beyond that they would expect you to step up to their commercial product offering.*)

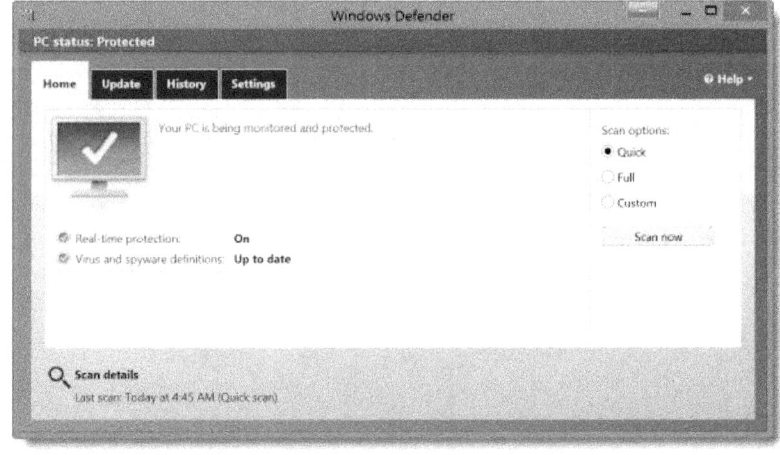

Screenshot of Microsoft® Windows Defender, Main Screen

Additional Malware Protection: I mentioned I prefer Microsoft® Windows Defender for my primary protection. Remember when I stated there were not any 100% perfect solutions for security? I do use some extra programs to complement the protection. A wonderful program at the top of my list is MalwareBytes Anti-Malware. (*The "paid" version of MalwareBytes now includes active virus protection and cannot coexist with other Anti-Virus products.*) This program, in the FREE form, nicely *coexists* with Microsoft® Windows Defender and scans with its own set of definitions and parameters. Moreover, it is not a "*burden*" on my computing resources, and is user-friendly. That is another key aspect for any security software: It MUST be user-friendly. If it requires an advanced degree in prime number factorization then most users will ignore it, get frustrated, and become unprotected. In addition, you can get the basic program FREE! There is that word again. By "*basic*", I mean you will have to manually run the program to have it scan your computer, once a week or whenever you decide your maintenance period should be. There is a small charge for the "*Professional*" version, (*at the time of this writing, $24.95 at* **www.MALWAREBYTES.org**), which would enable features such as automatic, full-time scanning. For business owners the price would be the same, with discounts for larger purchases.

RANDOM TECH TIP:

Just a quick tip regarding your online activities: Never assume what you do online is actually anonymous. Things often can, and do, get back to the source. Moreover, while we are on this subject, never post anything online or in an email that you would be afraid for your mother, or boss, to read! The unfortunate fact is that once you put the information out there it is possible for the world to see. Just ask some celebrities that thought their information was private!

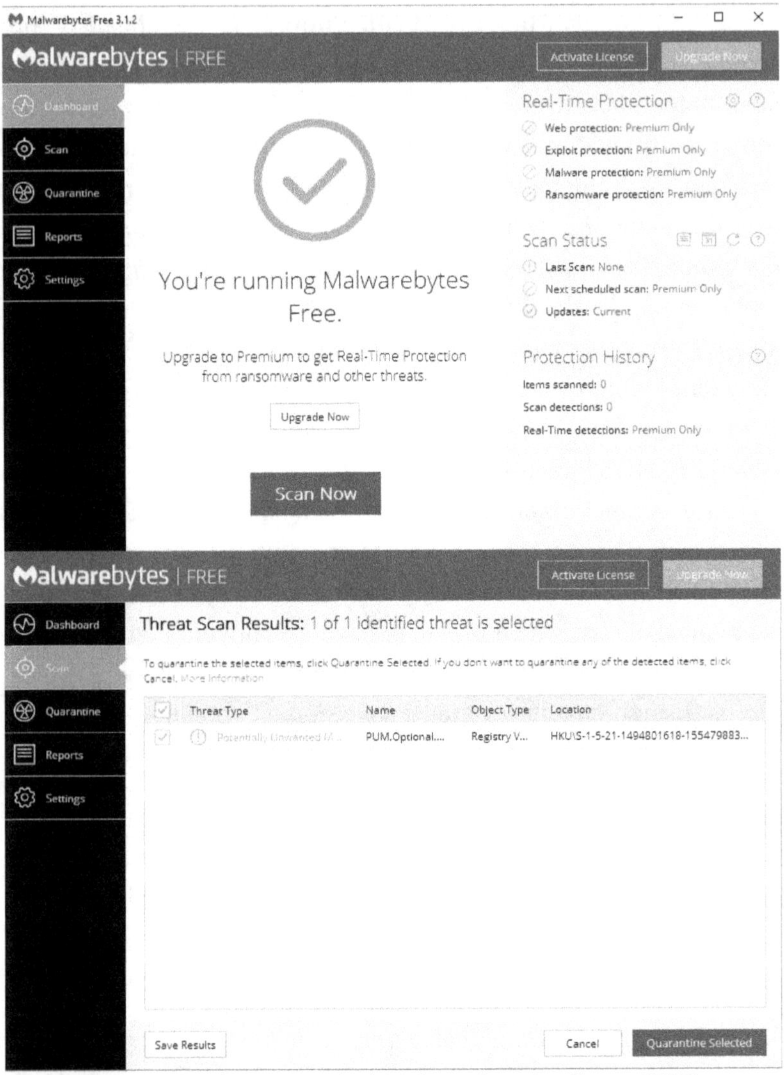

Sample of MalwareBytes Anti-Malware Scanning Screen (TOP)
Sample of MalwareBytes Anti-Malware Results Screen (BOTTOM)

What is the big deal with passwords?

The subject of passwords is painful to users and techies alike. Users think they are too much of a hassle. Techies keep asking you to make them complex, longer, do not reuse them, and do not write them down. Users retaliate by simply capitalizing the first letter of the word, adding a number to the end, and calling it a day. Everyone is happy, right? The happiest person may just be the criminal, as you just practically handed them access to your information. Your password may be the ONLY thing standing between your information and someone who would love to steal your identity! Just like I would be willing to wager that I could not simply ask any of you to give me your social security number or the PIN number to your bank account, I would HOPE you would be just as protective of your passwords. The sad truth is that many do not hold their passwords in such high esteem.

While developing this segment, I remembered a joke regarding passwords I heard a while back. I have NO idea who to credit the origin to, but hope they are flattered by their inclusion here. It told of the many ways passwords and underwear are similar:

- You should not leave them lying around.
- You should change them regularly.
- Best NOT to share with friends.
- They should leave a little mystery.
- When accidents happen, change IMMEDIATELY.

While originally designed as a joke, it illustrates some brilliant points. The ease with which many can obtain another person's password is almost scary. Combine the ease with the unfortunate fact that so many have the "*I don't have anything private on my computer*" syndrome and we end up with a real problem on our hands.

While items you have on your computer may at first glance not appear "*confidential*", let us take a moment to think of what is REALLY on that seemingly harmless computer:

- Pictures of family and friends
- Documents
- Email accounts
- Tax forms
- Receipts
- Friends and relatives contact information
- Travel plans

We could go on forever with this list. In my opinion, EVERYTHING on my computer is confidential! I consider it confidential because it is no one else's business but my own! Moreover, even in that short list could we see where there could be some dangers lurking? I do not want others having pictures of my kids, and do not want anyone contacting my family and friends.

One particular item on your computer often overlooked is the access to your email. This digital function has made its way into so many aspects of our daily lives. We can access it from virtually anywhere. We link accounts to our office computers, home computers, tablets, and smart phones. However, our distaste for passwords may mean access to our account has little, if any, protection. Could someone walk up, get onto your computer, and get to your email right now? How about on your phone?

If you answered "*yes*", please consider this scenario: Someone is upset with you and decides to use your email account to his or her advantage. While you are away, they send an email to your friend, spouse, or boss that is less than flattering. It came from "*you*", though, so accordingly the recipient is shocked, scared, or offended. In the extreme, this could lead to a "*resume-altering*" experience. At the minimum, it could result in embarrassment.

Many ask, "*Is there a real threat to someone "getting" my password?*" There indeed is a real and present threat. Most password hacking programs are fast and efficient and can break a majority of passwords in seconds or minutes. Their primary attack utilizes a dictionary scan. If you base your password on a dictionary word, including common misspellings and variations, it can be broken in seconds by easily available, often free, password hacking tools.

So how do we make a "*good*" password? For this, I would like to break out of the box a bit and introduce you to a different approach. Rather than a passWORD, let us instead consider a passPHRASE.

What is the difference? When most think of a password, they instantly think of a dictionary style of word: *Puppies, Kittens, Monkey, etc*. What I would like you to do is expand beyond the word and think instead of a phrase. With a passphrase, you open up the complexity while also increasing the ability to remember. Can we really make it longer, more complex, AND easier to remember? The answer is a most definite yes!

A passphrase can be a series of words which you can string together to make more complex, all while simply repeating them in your head to remember. Confused? Please do not be, dear reader. Many of us have sayings, lyrics, poems, or Bible verses running through our heads. The basic passphrase is simply making THOSE your password! For example, instead of "*Password123*", you could use "*NewOrleansSaintsWhoDat*!" While this may still appear as dictionary words, the length of the password, along with its run-on nature, make it much more difficult to compromise. This is therefore even more secure than even a random password, like "*v68P7qv0*", and a HECK of a lot easier to remember. Furthermore, any password you can remember without writing it down becomes that much more secure, in my opinion.

While we are in the neighborhood of passwords and passphrases let me take a moment to bring up one more access-related tidbit of information. Security questions are becoming a daily reality. You know the questions I am referring to in this statement:

- What is your mother's maiden name?
- What is your favorite food?
- What was your high school mascot?
- On what street did you grow up?

Let me point out a VERY important fact. There is no rule that you must answer these questions with truthful information. I know it is not nice to lie, but it is also not nice to steal people's information. This is a case where a little white lie can go a long way in protecting you. How much of this information could a total stranger gather about you from simply looking at your social networking profile?

To combat those who may have grown up with me, or those simply doing some social networking searches, from simply guessing my information and accessing my email or bank accounts, I have developed a simple and consistent set of alternate answers to these questions. Consider the following: (*NOT my actual answers*!)

- What is your Mother's maiden name? ~**Pizza**
- What is your Favorite food? ~**Smith**
- What was your High School mascot? ~**Purple**
- On what street did you grow up? ~**Bulldogs**

Again, as long as you stay consistent in your answers it would be virtually impossible for someone to take advantage of their knowledge regarding you. In this age of social networking, we answer these questions so often in our status updates and bio information.

- Mothers maiden names typically seen in their screen name
- Favorite foods mentioned when you go out to eat
- Bio shows your high school, and therefore the mascot
- Looking up your parents in your hometown shows street

∞ ∞ ∞ ∞ ∞

"The Internet is a telephone system
that's gotten uppity"
– Clifford Stoll

∞ ∞ ∞ ∞ ∞

† † † † †

RANDOM TECH TIP:

Email is a "best effort" system. There are multiple things going on in its transmittal that can delay or sometimes even stop its delivery. It is very similar to regular mail... most of the time you can count on it within a short timeframe, but every now and then, it is delayed. Email has to pass through sometimes hundreds of points to pass to its destination, and policies or issues at any of these systems can create delay. It is virtually impossible to trace the exact path an email takes. To trace its path would be the proverbial needle in the haystack. To deal with delays, email systems have a series of reactions and preventive measures. Each email is actually pre-programmed to keep retrying if it is held up at any point until it reaches its destination. This is why sometimes you receive them "late" by normal standards. If, however, they are delayed too long (typically 8 hours or more) they will send a notice back to their sender advising them of their delay. Then, they will keep trying, typically for two days. If after this time they are still unsuccessful, they will give up and send a notice to the sender before being erased somewhere within the Internet.

My Computer is SO SLOW!

"Man is a slow, sloppy and brilliant thinker; the machine is fast, accurate and stupid."
– William Kelly

What is likely the most common complaint about computers today? In my experience, it is "*my computer is so slow*". I was able to verify this as I put together material for this book. I posed the question out to friends and this was BY FAR the most common reply (*other than "will you please come fix my computer*?") We can agree it is a vague description and would be hard to diagnose from that statement. It would be like walking in to an auto mechanic and simply saying, "*My car isn't running right*".

A mechanic with less than stellar morals may hear that statement and envision a blank check. They can tell you anything, which also means charge you anything. This is not a good position to be in, right? It is the same with your computer. We need to dig a little further with some questions to ourselves. Some of these questions are "*difficult*" as it requires honesty. Here are a few questions to help find the underlying cause of the issue and get you back to work:

- What were you doing when you started noticing the problem?
- Have you installed, or updated, anything lately?
- Has anyone else been using your computer?

What were you doing when you started noticing the problem? The common answer I hear when I ask this question is "*I wasn't doing anything*?" Well, that should solve your problem. If you were not doing anything, then we OBVIOUSLY have an alien invasion of your computer causing such things to happen. The only solution will be to wrap the whole thing in tinfoil and call the folks at "*Unsolved Mysteries*".

While a malicious program running unseen in the background can cause it, the possibility remains that you may have been doing "*something.*" It is important to note that JUST because you were doing something does NOT imply you were doing something BAD. This is not a question intended to assign blame. I believe that sometimes the "*I wasn't doing anything*" answer is similar to the little kid with the hand caught in the cookie jar. Deny everything! It is OK to say what you were doing. It helps to get to the root of the problem and fix it.

- If on the Internet, the issue could be your browser.
- If checking your email, it could be your email client.
- If typing a letter, it could be your word processor program.
- If related to Internet use it could be your Internet connection.
- If simply staring at the screen it could be a rogue program running in the background.

Have you installed anything lately? It could mean that new little program you saw in that spam message may not be the most stable. It could also mean the last update applied to your trusty, faithful program broke more than it fixed. Either way it is a valuable bit of knowledge to have when troubleshooting an issue. Operating systems, as well as other programs, update regularly. Most of the time these install and everything continues working just fine. Unfortunately, sometimes there are issues. I will not go into heavy detail on this yet, as this is a topic for a later section of the book (*Technology Care and Feeding*).

∞ ∞ ∞ ∞ ∞

"Describing the Internet as the Network of Networks is like calling the Space Shuttle a thing that flies"

– John Lester

∞ ∞ ∞ ∞ ∞

Has anyone else been using your computer? There may be occasions when someone may be sharing your computer. Spouse, kids, relatives, pets. Who knows what they may have done while you were not looking. When asking this question a common reply I receive, besides the "*I wasn't doing ANYTHING*", is that their child used it for completing some homework. Once I dig in a bit deeper, I see that they used it for homework while also installing an application to chat with their friends while also streaming in the latest songs from an online provider. This opens up a giant can of "*What in the World*". This also gives good direction for troubleshooting the issue(s). A fix could be as simple as uninstalling the last program(s) installed.

It would be simpler to build a 4-lane bridge to Hawaii than to attempt here to explain all the possible scenarios and solutions to what may be making your computer act up. What each of us does with our computers is very different. As such, troubleshooting is different for different scenarios. However, there are some general tips to follow to relieve many common issues:

- Close any open programs.
- Reboot your computer.
- Reboot your Internet connection equipment.

Close any open programs. It could be something has gone a little crooked for a program you are running. A nice place to start would be to close what you have opened, such as your email or Internet client, wait a few moments (*30-seconds to one minute is typically adequate*) and then re-open it. You could find things start responding considerably faster. I have seen many cases where having too many things opened at once can cause performance issues as well. This is especially true for older hardware.

Reboot your computer. The good old answer so many tech support folks love to spring on you. Nevertheless, it really IS great advice, as it can fix many issues. Still, many people wonder why this is true. The short answer is there is no one-size-fits-all answer as to WHY it works. However, I can offer an analogy to explain what happens and perhaps make it a bit clearer:

When your computer is running, it stores a large portion of its operational information in RAM (*Random Access Memory*), which is the "*working*" memory for your computer. Think of this memory area as the top of your desk. With every program you open, web site you visit, and email you read you place more information on top of your desk. You may close these items but they still sit on your desk, likely "*covered*" by the next web site or email. As you can imagine, things can build up and become cluttered. Rebooting your computer is the same as removing every item off your desk, again revealing that beautiful wood grain beneath, and getting a fresh start to your day. You are not sure which lingering item was creating the issue, but you do feel relieved as you can once again return to working normally. The important thing is that things work. Your friendly neighborhood scientist may argue it was solar flares charging to 1.21 Jiggawatts while attempting to reach 88 MPH on the Internet, but the reality is you simply cleaned your desk and started fresh!

It is important to note there IS a difference between "*RESTART*" and "*REBOOT*". When you RESTART, your computer never actually turns all the way off. It is what we nerds refer to as a "*warm reboot.*" When you perform a warm reboot, it *may* not fully clear out everything, as the computer never truly loses power to all parts. We typically accomplish this by selecting "*Restart*" from the shutdown options on your computer. While many times a restart may cure your issues, it is not a 100% cure.

WARNING: *Be sure to save any documents, etc. that you are working on! You would not want to lose something important in this process!*

REBOOT means to completely turn off your computer. This removes all power to the computer's memory, drives, and such. This is what we nerds refer to as a "*cold boot.*" A reboot will fully accomplish our goal of "*cleaning your desk*" and result in a nice fresh start for your computer. In my opinion, when you are looking to clear an issue this is the best option. I advise users to leave their computer off at least one minute prior to turning power back on.

Like I mentioned earlier, be sure you SAVE any items you are working on before restarting or rebooting.

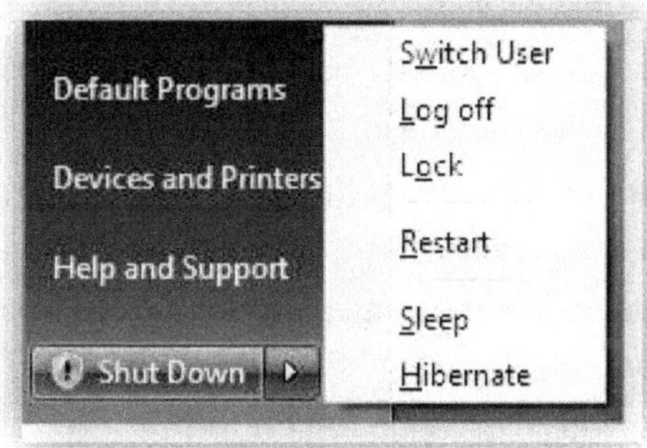

Sample of Windows® Shutdown Options
Click SHUTDOWN to Reboot

Reboot your Internet connection equipment. Although we will cover your home network more extensively in an upcoming chapter, I would be letting you down to not mention this now. Many households today have high-speed Internet access, through your phone, cable, or satellite provider. These connections involve a special piece of equipment known as a **modem** to connect to the provider and provide your Internet connection. Like your computer, these items can benefit from a restart.

With so many options available for home networking, to actually tell you WHICH item(s) in your home I am talking about would be hard to do. What I can tell you are a few tips to identify the equipment. We will discuss this in further detail in the next chapter:

- Your modem was likely sent to you, and possibly installed, by your service provider. If you currently have DSL, U-Verse, or Fiber services this would likely be through your telephone company. If you have a cable modem this would, of course, be through your cable provider.
- Depending on the type of connection, your modem will have either a regular telephone line or a larger, round TV cable plugged into it:

Coax Cable (LEFT), common to Cable Modems
RJ-11 Cable (RIGHT), common to DSL Modems

- Often the modem may NOT have a power button to turn the unit on or off, so the only way to reboot the unit will be to remove the power cable. Wait one minute before plugging it back in to allow it time to reset.

∞ ∞ ∞ ∞ ∞

"One good reason why computers can do more work than people is that they never have to stop and answer the phone."

– Unknown

∞ ∞ ∞ ∞ ∞

If you have wireless capabilities this may either be built-in to the service provider's modem, or a separate unit connected to the modem, commonly by a cable that looks like an overgrown telephone cable. This will need a rebooting as well, if it exists, following the same one-minute recommendation.

RJ45 Cable, commonly Used to Connect Network Equipment

- If you have two units, as described above, I recommend turning on the modem first and allowing it to come back to service before turning on any additional equipment such as the wireless unit. I recommend waiting at least one minute between plugging in each device to allow them to return to service properly.

Once you have restarted the associated equipment you should see things settle out and behave better. Just like us, a fresh start can be a good thing. What if your problem persists? Well…then, it may be time to dig a bit deeper. Unfortunately, not all issues vanish after a reboot. Some require application of advanced nerd expertise. As much as I would love to, I do not believe there is enough ink to make a comprehensive troubleshooting book for your computer for every situation. There are simply too many variables. Nevertheless, keep reading and I will continue to share hints and tips to help you get through the more common issues between technology and you!

Networking at the House

"They've finally come up with the perfect office computer. If it makes a mistake, it blames another computer."
– Milton Berle

...or should this chapter be titled
"How to Sort the Mass of Wires Behind Your Desk"?

If there is ONE technology item that aggravates my wife it is the mass of wires "*required*" to bring this technology demon to life. Even in this age of "*wireless*", we still see the electronic octopus dominate the backs of desks nationwide. With all of our untethered devices, we still need wires to route signals and power to our equipment. So what do they all go? In this chapter, we will work to make sense of the mess and bring about an understanding of how to make the most of your home network. We will discuss things to look for when choosing a wireless router along with some basic troubleshooting tips. We will also reveal some best practices to make sure we get the most from your equipment purchases.

The core of your home (*or office*) network is your modem. Your ISP or "*Internet Service Provider*" usually provides this device. Simply put, these are the folks to whom you pay monthly fees for access to the Internet. The most common methods being a cable modem (*your cable provider*) or DSL (*Digital Subscriber Line, from the phone company*). You may also have satellite service, U-Verse or Fiber Internet offerings. We could argue at length over which is better. In the end, we would arrive no closer to the truth. They each have advantages and disadvantages, so go with the one that offers you the best overall experience and comfort level (*price and customer service come to mind*). After all, you will end up in the same place…on the Internet.

∞ ∞ ∞ ∞ ∞

"People think computers will keep them from making mistakes.
They're wrong. With computers you make mistakes faster."
– Adam Osborne

∞ ∞ ∞ ∞ ∞

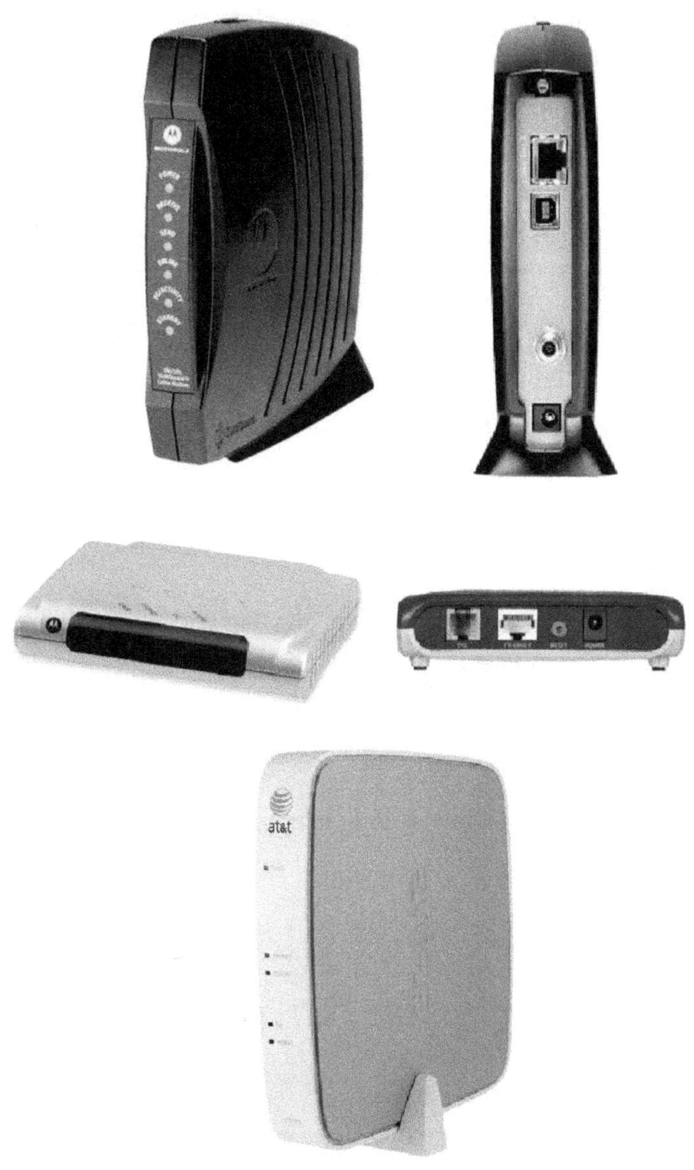

Common Cable Modem, Front and Rear (TOP)
Common DSL Modem, Front and Rear (MIDDLE)
Common DSL Modem with Built-In Wireless Router (BOTTOM)

Things that happen beyond the modem can be as varied as the equipment we use to undertake the task. There does not seem to be a one-size-fits-all scenario, but there are some common features:

- Some have just one computer and the buck stops there.
- Some have wireless, expanding the amount of connected devices exponentially.
- Some have extended wiring throughout their home (*or office*) connecting many devices.
- Beyond computers, some connect printers to share throughout their network.
- Televisions and gaming consoles connect, providing rich multimedia experiences.
- Even storage devices get on the network, providing shared storage to all users.
- As homes get "*smarter*" we see connected thermostats, cameras, and even personal assistants, like the Amazon Echo®.
- Moreover, some have a combination of the above items!

† † † † †

RANDOM TECH TIP:

If you find yourself having to reboot too often (intentionally vague…we would all define "too often" differently; personally, I define "too often" as multiple times daily) it could be an indication of a bigger problem and may be worthwhile to have the issue looked into further.

If you have one device connected, you have a rather simple scenario, right? Not quite. There are still quite a few things one should consider, no matter how big or small the network. Although the single-device user may not have as much wire hiding behind their desk they still need to know what is going on for everything to work. We will first start explaining what is going on with this single-device scenario, growing towards a more complex network consisting of multiple devices.

We already mentioned that to access the Internet your service provider would provide you with a modem, likely at a cost. The TYPE (*cable, DSL, etc.*) of access you have is not truly important at this point, as the ultimate goal is, again, the same. We want on the web! To accomplish this, your modem essentially becomes a door, or "*gateway*" into the Internet. Whether you have one or one thousand devices, they count on this gateway to lead them out onto the Internet.

This gateway created by the modem is actually intelligent. It does a lot more than simply open the door whenever you need "*out*". One function of your gateway is to give you directions. It does not want to send you out unprepared! When you request a web site, or other item, from the Internet, your gateway helps to determine the best route for "*you*" to take in order to retrieve the information. This is a setting within the gateway known as DNS, or Domain Name System. Without getting too nerdy, this is a function on your network that takes the web site name you enter into your computers browser, such as **www.MPBOnline.org**, and converts it into a series of numbers (*currently 54.225.206.152*). It turns that name into an "*address*" the computer can understand. By performing this step, your gateway makes web browsing much more convenient. It is MUCH easier for us to remember names than a series of numbers. Why is this nerdy tidbit important? We will see in the later chapter on surfing safely that we can use this function to keep our browsing safe from harmful and offensive sites. This will also assist us when troubleshooting Internet issues, as we will see soon.

Your gateway can do even more. They can actually stand guard for you, keeping unwanted "*visitors*" from entering. In this role, it becomes a **hardware firewall**. When unsolicited traffic attempts to enter from the outside world, it compares the visitor's information against a set of rules to determine if they are allowed in or denied access. In most residential situations, this inspection would result in a denial of the uninvited request. In a corporate setting this may change slightly, as you may have web sites internally that you wish public visitors to be able to view.

If you have but one device, your network MAY stop here. Increasingly we are seeing homes with many more connected devices, wired and wireless. Multiple computers for different family members, tablets, smart phones, gaming consoles, and even televisions seem to be clamoring for attention from the Internet. As such, many have seen the need to expand their core equipment to add increased functionality. One of the more popular additions is that of the wireless router. While this function may be built-in to your modem, especially in the case of DSL, more commonly it is an added-on piece of consumer electronics picked up at your local store. Moreover, like so many things these days, it can have many features depending on the model chosen. Rather than get into all the bells and whistles, we will focus on the general features for the sake of simplicity and minimal nerdism.

∞ ∞ ∞ ∞ ∞

"Computers can now keep a man's every transgression recorded in a permanent memory bank, duplicating with complex programming and intricate wiring a feat his wife handles quite well without fuss or fanfare."

– Lane Olinghouse

∞ ∞ ∞ ∞ ∞

Going forward we will be focused on wireless connections, but please know that many of these same tips and concerns affect wired connections as well. The primary exception being that in a wired connection someone must physically connect to access your systems. Reality today is that wireless is practically "*expected*", so we will turn our attentions to the airwaves.

DISCLAIMER: *Your added-on device(s) may or may not have some of the described features. Consult your owner's manual or the manufacturer's website for details.*

RANDOM TECH TIP:

*When you start receiving delivery failure notices for emails you did not send it is called "backscatter". Authors of spam and viruses wish to make their messages appear to originate from a legitimate source to fool recipients into opening the message so they often use web-crawling software to scan web comment postings, message boards, and web pages for legitimate email addresses. The "bad guy" pretends to be you (*called "spoofing"*) and sends random messages in bulk to sometimes thousands of email addresses. When this happens, YOU end up getting multiple delivery failure notices, or even sometimes replies from these people receiving this mass junk mail.*

For more information on this issue:

http://en.wikipedia.org/wiki/Backscatter_(email)

Choosing a Wireless Router

When you are standing in the wireless router section of your favorite store, staring at the boxes may seem like you are staring into a kaleidoscope. All of the terms seem pretty, but you are not sure what to make of them. Let's see if we can decode the boxes together (*now, if I can just get one of you to help me understand nutrition labels, and why breaking a cookie in half doesn't let all the calories fall out*...). Below are some terms and features you may see listed, along with a brief description of their meanings:

- **Dual Band** – This lets you know that the wireless router can operate on the 2.4GHz range as well as the 5GHz range for wireless communications. We will discuss this in more detail later in this chapter.

- **Ethernet Ports** – These are the additional spots available to PHYSICALLY plug wired network devices, such as computers or printers. The cabling for it looks like an overgrown telephone cable (*nerd terminology: this is known as an RJ-45 plug, as shown previous page*).

Ethernet Ports (*numbered* 1-4) on Rear of Wireless Router

- **IPv6** – This is a new structure for assigning numeric addresses to networked devices. It is nice to have the support, though in my experience it is not popular…yet.

- **Gigabit** – This refers to the speed for WIRED connections.

- **Wireless N & AC** – Wireless N and AC are technologies. They consist of advanced features affecting speed, distance, and performance. You will see the "*type*" of wireless referred to by the letters A, B, G, N and AC. Without getting into the nerdy specifics, know that AC is the latest and greatest, and considered the fastest.

- **Extended Range/MIMO** – Refers to technology the manufacturer uses to extend the range of the device, typically involving the use of multiple antennas. These antennas can be internal to the unit, so you may not "*see*" them.

- **Ready Share/USB Storage Port** – This is a feature to allow you to connect an external storage drive to permit sharing storage with multiple users on your network. We will discuss this further in this chapter.

- **QoS** – Stands for "*Quality of Service*". This advanced feature of some units allows you to optimize your network for things such as streaming video from online services.

∞ ∞ ∞ ∞ ∞

"Before you become too entranced with gorgeous gadgets and mesmerizing video displays,
let me remind you that information is not knowledge,
knowledge is not wisdom, and wisdom is not foresight.
Each grows out of the other, and we need them all."
– Arthur C. Clarke

∞ ∞ ∞ ∞ ∞

Common Wireless Routers

∞ ∞ ∞ ∞ ∞

"To err is human, but to really foul things up you need a computer."

– Paul Ehrlich

∞ ∞ ∞ ∞ ∞

While the addition of wireless to our network does a GREAT job in minimizing those wires my wife despises so much, it adds a few more concerns. Now, mind you, these do not concern HER…she just wants it to work. However, they DO concern ME:

- Environmental factors affect performance.
- The more devices, the more slowdown you may experience.
- With wireless, your network extends beyond your walls.

Environmental factors affect performance. This is an area in dealing with wireless networks often overlooked. WHERE you position your wireless device can significantly affect its performance. There are devices and areas in your home or offices that are not wireless-friendly and can rob performance. For example:

- Do you have cordless phones? Many cordless phones, and baby monitors, operate in the 2.4GHz range, the same as older wireless network devices for the home.
- How close is the device to your kitchen or laundry room? Appliances like your microwave in these rooms use a lot of electricity and generate interference to wireless signals.
- Are there fluorescent lights nearby? This applies more to office settings. Fluorescent lights generate a lot of electrical "*noise*" that can degrade performance. NOTE: This applies to WIRED connections as well! Never allow network cables to run across fluorescent light fixtures!
- Distance between the wireless access point and your device can affect performance.
- Other neighbors with their wireless signals can cause "*congestion*" on the airwaves.

While some of these can be resolved by simply relocating your wireless router, we can address others with technology. Many new wireless routers have the added ability to function in the 5GHz range to reduce the interference. Not all equipment will accept this new range, however. Older equipment is likely limited to the "*original*" 2.4GHz range. For this reason, most 5GHz routers also include options to run in the older "*legacy*" 2.4GHz range for compatibility. This is referred to as Dual-Band functionality.

The more devices, the more slowdown you may experience. Think of your Internet connection as a water pipe. You can only push so much through it before something starts to back up. Likewise, you only have SO much space to give to your network devices before something has to wait. This could mean that if your spouse is busy uploading videos to their social media site you could see connection issues while you are playing an online video game. (*YES, sadly, this happened to me*.) When interference like this happens you will notice things seem to stop or slow down. In this day and age where so many are streaming movies directly into their home you would see this as a pause in the movie while it attempts to "*catch up*", fighting for space in your "*pipe*." If you are downloading information, you may see a slowdown in the download speed, or even a complete failure. Essentially, any activity reaching out through your gateway competes for space in the pipe. Imagine a crowded highway when you are late for work. Yep, that is what this congestion feels like to your computer!

∞ ∞ ∞ ∞ ∞

"A computer will do what you tell it to do,
but that may be much different from what you had in mind."
– Joseph Weizenbaum

∞ ∞ ∞ ∞ ∞

With wireless, your network extends beyond your walls. What do I mean by saying your network now extends beyond your walls? We must remember that much like radio station signals, wireless can travel well beyond your private little area. Therefore, chances are high that your neighbor(s) can also see your wireless network signal. Moreover, if they can see it they could connect to it. Some of you may be thinking, "*So what if they borrow a little browsing time*?" Well, let me offer a couple of thoughts:

- What if your neighbor decides to use YOUR connection for downloading illegal materials? Network/Internet traffic is traceable by your ISP, so who do you think will get the knock at the door from the police?
- If your ISP provides limits to how much Internet bandwidth you can use per month, a neighbor "*borrowing*" your signal can cost you money in overages.
- If a neighbor gets onto your network, they can potentially see all your devices, and all the information on those devices (*Emails, documents, pictures, etc*.).

Wireless routers can add many advanced features as well:

- Ability to restrict browsing times
- Capability to apply parental controls
- Connecting external hard drives for shared storage/backups

Ability to restrict browsing times. Sometimes as a matter of keeping the peace, it may be in your best interest to restrict available Internet surf times. I have seen parents want to apply "*quiet time*" after a certain hour, or during homework time, to prevent distractions or ensure their kids follow bedtimes. Many modern routers allow you to restrict the time(s) devices can access the Internet.

Capability to apply parental controls. With so many inappropriate sites out there today, many parents choose to apply filtering techniques to keep some of the worst offenders out of their homes and off their children's devices. This can have the added benefit of keeping a good bit of malicious software out as well. We will cover these features in-depth in the next chapter.

Connecting external hard drives for shared storage/backups. I have seen this feature used more and more. External drives in varying sizes have become increasingly cheaper. This opens up a huge potential in our connected world. Instead of attaching these drives to individual computers for single-use, many routers will allow you to connect the drives directly to themselves. In this case, you can then share the storage among multiple devices. With the multitude of manufacturers out there, the precise steps to accomplish this may vary. If your device is capable, I would suggest reading up on your device and exploring this potentially unused but beneficial feature.

Common External Hard Drives

Wireless Connection Security

We could write an entire novel on this subject alone! So often overlooked is the wireless security segment! In our quest for EASY we open ourselves up to abuse. Let us discuss a few wireless rumors:

Wireless passwords are hard to set:

- FALSE: wireless passwords are easy to set on home wireless routers. The hardest thing to do may simply be to come up with the password itself!

I will have to enter this password on each device I want to connect:

- TRUE: you will have to enter your password when you INITIALLY connect a device. From that point on your device (*laptop, tablet, etc.*) should remember the password and auto-magically connect whenever it next sees your wireless network in range. It will also remember others and reconnect when you return, such as at a relative's home.

No one would really connect to my network, and surely they wouldn't do anything BAD if they did!

- FOR SALE: Oceanfront property; beautiful views, sandy beaches, winds off the ocean. Come see in the Great Smokey Mountains. Included with purchase price is the Brooklyn Bridge. Serious offers only. Deposit required before crossing the bridge. Call 555-1212.

When you go into your wireless router to set wireless security settings you are presented with a few common options. As the steps to get into your wireless router vary, you would need to consult your documentation for specifics. Depending on your model, there may be others, but these are your common "*core*" components:

- Name, or SSID (*Service Set Identifier*)
- Security Type/Encryption (*None, WEP, WPA, WPA2, etc.*)
- Desired Password/Key
- Guest Network Settings

Name, or SSID (*Service Set Identifier*). This will be the name for your wireless network. I HIGHLY suggest changing this from whatever default name your device has assigned to something with meaning to you. I would not, however, recommend using your name. It makes it easy to identify by some would-be crook. If they see "*Smith House*", they know exactly where it leads. Using something easy to identify is considered bad security practice. I have seen people get VERY creative with these names, with SSID's such as "*FBI Surveillance Van*", "*Howdy Neighbor*", and "*Beware of Dog*". Potential names are limited only by your creativity.

You should be sure to change the wireless network name, password, and any other passwords that came from the factory when installing your wireless router. Why is this important? There are lists available on the Internet that reveal the default security settings for every device on the market. By leaving your settings as these factory settings, it is the same as leaving your car keys in the door. You would be inviting a person with criminal intentions into taking advantage.

Security Type/Encryption, typically: None, WEP, WPA, WPA2, etc. The possible entries of this type vary based on the age and model of your device. Even so, there are some basic guidelines you should follow to secure your connection. If it is an option, choose WPA2. In descending, and weakening order then would be WPA, WEP, and finally, none. WPA2 will offer you the best protection available in a consumer environment.

Now we could ramble on about encryption, but let us just put it simply: Encryption is the scrambling of data to make it unreadable except by the intended recipient. In other words, it scrambles your computer activity across the air between your computer and the wireless access point. Anyone therefore intercepting, (*referred to as "sniffing*"), the encrypted traffic would see unreadable, garbled text.

Desired Password/Key. In this area, you would enter the password to connect to your wireless network. Please refer back to the "*Pass Phrase*" discussion the "*Computer inSecurity*" section for ideas on creating a good phrase to protect your network.

Guest Network Settings. New routers have an option for you to enable a guest network. Why, you ask? When you have a guest come to your home and they need to access Internet, on perhaps their laptop, you can provide them access to this guest network instead of your private network. Just think…if their computer has malicious software on it and you put them on your private network with your computers they could actually infect YOU as well! The guest network allows them access while preventing them from "*seeing*" your equipment. In addition, by setting this guest network to a different password, you prevent revealing your private password.

Troubleshooting Network Issues

It seems inevitable we will experience issues in our connected world. It can be frustrating, and at times difficult to diagnose. Troubleshooting network issues require we first ask ourselves a few questions. Otherwise, we will spend a lot of time running in the wrong direction when all we really want to do is get back to work, or play.

- Are there any orange lights on your network devices, such as the modem?
- Are you experiencing the issue on one device, or on multiple connected devices?
- If an Internet browsing issue, is the issue with a specific web site, or with all web sites?

Are there any orange lights on your network devices, such as the modem? Let us always start with something simple. Without knowing the specifics of YOUR model of equipment, I can say that in 95.342% (*non-scientific*) of cases, the illuminated lights on your modem should be green. Why not 100%? Well…in technology there are no absolutes. The BEST thing to do is to make a note of what NORMAL is to your equipment when things are working. I have seen people, smartly, make a written note on a Post-It® and stuck it to the device indicating what NORMAL is for your configuration. That way, when things seem wrong, you will have something to compare. They record, or draw, what the lights normally look like, such as if they are blinking, off, green, etc. I have provided an easy to use area to record this information on the last page of this book.

If you do notice things do not appear "*normal*" it would be best, and simplest, to restart the equipment. We discussed this earlier and would be applicable here as well. If the restart does not help at this point, you may need to call your provider. **Please note:** Your ISP will NOT troubleshoot problems for you on equipment YOU bought and installed, such as a wireless router from the store. Their involvement would stop at the equipment THEY provided, which is likely only the modem itself. For these other devices, a call to the manufacturer may be in order.

Are you experiencing the issue on one device, or on multiple connected devices? If you are experiencing problems with just one computer, it is very likely the issue lies within that one computer. Primarily we should employ our tried and true first step of technology troubleshooting: *reboot your computer*. We discussed it in-depth in the first chapter, and we will probably see it again as we continue. This quick and easy step can save a lot of time, and money, when dealing with technology. Besides, if you call tech support the first thing they will ask you to do will be to reboot your equipment, so you might as well try it first. (*A reboot does NOT involve kicking your computer with boots on, no matter how much you feel like doing such*).

RANDOM TECH TIP:

When your Internet connection is down, finding the number to customer service may be difficult! Make sure you record the support numbers somewhere ahead of time for such emergencies. As a suggestion, you could write them on the back page of this book to keep it them handy! (If you are reading a digital copy, please feel free to attach a Post-It® to your e-reader!).

If the issue is affecting more than one device, we could presume the issue is with a "*common*" component. In most cases, this would be the core network equipment, such as the modem or wireless router. The quickest and simplest thing to do first would be a reboot of the equipment. Even if there are no lights on besides the "*normal*", an issue could be underlying that is creating the problem. A simple reboot of equipment never hurts. If this reboot fails to resolve the issue it is likely an ISP issue. A phone call to them would be the next best step, as they can quickly tell you if they are experiencing any service issues.

If an Internet browsing issue, is the issue with a specific web site, or with all web sites? This is the first question I ask when someone reports to me they are having an issue with a web site. They try an address they have gone to many times and now it does not work. In their mind, the Internet is not working. Truth is it could be the site is merely having an issue.

I would ask the user to try one or two other sites at this point. Sites like **www.GOOGLE.com** and **www.MICROSOFT.com** are my typical requests. These large sites are rarely, if ever, taken offline. If unable to reach these sites, my troubleshooting would lead me to believe I had a much bigger problem on my hands.

∞ ∞ ∞ ∞ ∞

"Give a person a fish and you feed them for a day; teach that person to use the Internet and they won't bother you for weeks"

– Unknown

∞ ∞ ∞ ∞ ∞

If you are able to reach other sites, but you continue to have problems reaching a web site you need, there are free services available to help you. If you would like to double-check if the site is in error, you can use this handy web site to check the status of the web site: **www.DOWNFOREVERYONEORJUSTME.com**. Long name, but very self-explanatory. Type in the address of the site you are trying to reach and it will tell you if others are experiencing issues as well. Short, sweet, simple and free! If the web site is shown as down for everyone you simply must wait for the issue to be corrected. If your research shows the site IS functioning properly for others and you can still not successfully reach it, this *could* be an indication of a malicious software infection. Many times an infected computer will attempt to disrupt your web browsing to keep you from cleaning off the infection. This may be a good time to run some of the security software discussed in Chapter 1. Of course, a quick reboot of everything would still be a good idea, as it is free.

If I ask the user to attempt to open additional sites and they are successful in reaching the sites, I can presume the issue lies with the site they are trying to reach. This could be due to a couple of things, but most commonly, it is simply a web site error. The first, and simplest, thing to do is to double-check the name of the site you are attempting to reach. I cannot tell you how many times I have seen a simple typographical error being the cause of a web-browsing breakdown. Another possibility is the site could be performing updates or experiencing issues. In this case, I would simply advise someone to try again later, much like if you were calling a business and found the phone line to be busy.

So far, the main "*troubleshooting*" advice I have shared has been to reboot your equipment. Unbelievably this works in 94.865% (*non-scientific*) of cases, in my opinion. What if we dig a little deeper? We will discuss troubleshooting common PC issues more in-depth in a later section of this wonderful book. I will mention a few tips on the upcoming pages that may assist in resolving your network issues.

Do you think your network connection is slow? While not an exact science, there are sites that can help you test your connection to see if things are performing OK. One common type of test is the Internet speed test. These online sites check your speed uploading TO the Internet as well as downloading FROM the Internet. A popular provider of this free analysis is **www.SPEAKEASY.net/speedtest**. You would go to the site and simply pick a city closest to you to evaluate the approximate speeds you are receiving through your ISP. You should make sure you, and no one else on your network, is using the Internet at the time to prevent interference. It is IMPORTANT to note these are NOT exact scientific tools, and are simply a guideline.

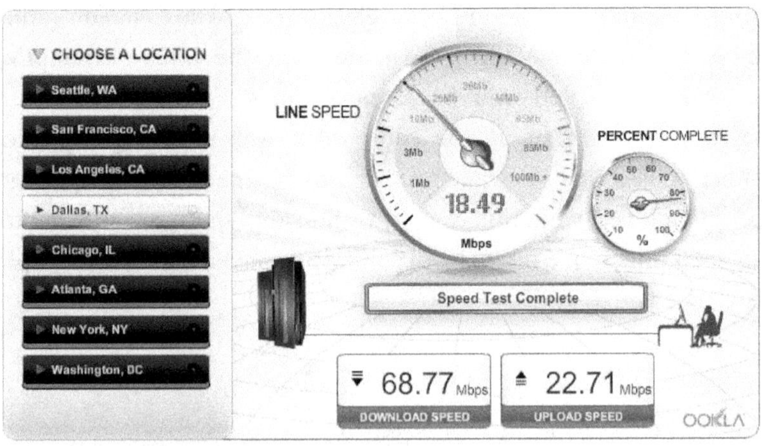

Sample Screenshot from Speakeasy Speed Test

NOTE: In many residential situations the download (*or "from"*) speed is much faster than the upload (*or "to"*) speed. The reality is that not many, if any, of us run things from the house that would require us to send information TO the Internet. For example, I would venture a guess that barely any of you reading this have a web server running AT your house. As such, your ISP charges you less than a business customer who DOES need this capability, since you do not require the speed in that direction. Download speed is where most residential customers need performance, and that is therefore where the ISP concentrates their efforts, and costs, in providing you Internet service.

You can't get ANY web site to open up? Remember before when I mentioned that tidbit about DNS converting the names we enter into our browser into numbers a computer can understand? Now we can put that knowledge to work for some troubleshooting. If entering names in your browser still fails to bring you back any web site results, we can test that DNS is working by plugging in numbers instead. Instead of the name, enter the following numbers and you should see their respective sites appear:

- Google®: http://74.125.140.99
- Microsoft®: http://64.4.11.37
- Amazon®: http://72.21.214.128

If entering those numbers results in you successfully seeing the page, you have a failure in DNS. If entering those numbers fails to display their respective web sites, then you likely have a larger connectivity issue. I presume at this point you have already restarted your equipment, so a call to your ISP may be in order in either of these cases, as in most cases your ISP controls this service. You should report your findings from this step to the technician.

My wireless coverage is weak in some areas of my house. WHERE you position your wireless router can significantly affect its effectiveness. Placement helps maximize coverage and strength throughout your home. Here are a few tips to get the most of your wireless network:

- Place the wireless router as close to the middle of the house as possible. This will help ensure coverage from side to side, and front to back.

- If your wireless router has an external antenna (*some antenna are inside and non-adjustable*), position it vertically to maximize the signal broadcast.

- Elevate your wireless router. Sitting the device on a floor is not a good spot. Being on a desktop will improve reception and broadcast.

- Avoid placing the wireless router on a metal surface, such as a desk or filing cabinet, if possible. Metal obstructs signals from transmitting and returning to the device.

† † † † †

RANDOM TECH TIP:

Get to know Google®, or your favorite search engine. Many issues you are experiencing can be solved by an online search. I have long believed there are rarely any UNIQUE issues that you alone are experiencing. Someone else has likely run into the same issue. Simply go to your favorite search engine and type your question!

† † † † †

Safely Surfing the World Wide Web

"The Internet is the most important single development in the history of human communication since the invention of call waiting."
– Dave Berry

Perhaps we could better describe it as the "*World Wild West.*" It is like Billy the Kid meets Jesse James, with a dash of The James Gang. If we could only find Wyatt Earp and clean the place up a little! Simply getting on the Internet exposes you to many dangers lurking around every click. This section will look at ways to protect you from these desperados.

Internet Monitoring

Internet monitoring is quite possibly the most despised feature I have in place at work. Users dread seeing the screen pop up when trying to browse the Internet stating they cannot view the page because of some rule or policy. How dare someone say the site is not work related! How do they know the content of the site, after all?

Internet monitoring is very similar to Anti-Malware software. It relies on definitions and patterns to classify the billions of web sites in existence today. As you could imagine, this is an ever-changing landscape, so it is not a perfect system. In my experience, though, it has a good record of accomplishment.

Like Anti-Malware software, sometimes the monitoring software will get it wrong. It does happen, and there have to be adjustments made to suit your environment and expectations. There is no way for a product to predict YOUR web habits, YOUR business requirements, or what YOU consider offensive. In reality, you must customize system settings to fit your needs. In a corporate environment, this frequently consists of notifying management and the IT department of your business requirements. At home, it is up to you to make adjustments.

Like Anti-Malware, Internet monitoring makes use of definition files to categorize web sites and make filtering much easier. We take it for granted when sites containing pornography, hate, or discriminatory content are blocked. There are also other categories with which you may not be as familiar and these can be prevented from making it onto your home or business computer. For example:

- P2P/File Sharing
- Parked Domains
- Proxy/Anonymizer

P2P/File Sharing. P2P, which stands for "*Peer to Peer*" is a method for sharing of files across the Internet between hundreds or thousands of people. While some use this method for legitimate activity, the unpleasant reality is that P2P has been branded as, and become a haven for, the trade in illegal software, music, videos, etc. If you ARE using this for legitimate purpose then you obviously know what the program is, how to operate it safely, and how to compensate for the settings required. If you are reading this and scratching your head as to what in the world this means, we can safely assume your best option would be to block these sites from being accessible.

On the darker side of the Internet, P2P sites and software exist for the trading of illegal material. Since the source of these items may be less-than-legitimate, the possibility of the files being malicious increases exponentially. I cannot literally count the number of computers I have had to repair which have been literally decimated due to someone installing and downloading software using P2P sites.

Parked Domains. When I mention the term "*Parked Domains*", I typically see a blank and distant stare in people's eyes. What is a parked domain? A parked domain is a web site owned though currently has no relative information on the site. People do this for many reasons, such as:

- Reserve the site name for future use
- Protect the name from being bought by someone else
- Cybersquatting

Cybersquatting is a potentially dangerous issue for our Internet safety. These cyber squatters purchase the site with a malicious intent. These sites may have been ones previously owned by a legitimate company. They can also be sites consisting of legitimate site misspellings. Malicious users will then populate these parked domains with their own malicious software and advertisements. The worst of the offenders will populate the site with pornographic material. One popular past example was ***www.whitehouse.COM***. The REAL site address is ***www.whitehouse.GOV***. The incorrect site was previously a pornographic site. One could imagine the shock of parents and students researching information for a civics class project!

∞ ∞ ∞ ∞ ∞

"We are stuck with technology when what we really want is just stuff that works"

– Douglas Adams

∞ ∞ ∞ ∞ ∞

Proxy/Anonymizer. Proxy servers are servers commonly used to bypass attempts to filter and block users from visiting "*bad*" sites. For example, a student would send their request for a restricted site "*through*" a proxy service to get THE PROXY to retrieve and send the information to them. This would trick the filtering service from seeing the web request to the restricted site.

- Student types **www.*NaughtySite*.com** in browser
- School security software BLOCKS access to site
- Student visits **www.*PROXY*.com** and request site
- School security software sees student website request as **www.*PROXY*.com** and is prevented from seeing the TRUE requested address of **www.*NaughtySite*.com**
- Student is now able to view restricted content

An Anonymizer is a service on the Internet used to hide the identity of the user. This essentially becomes a shield between your computer and the Internet. While on the surface, this may seem like a good idea in a world of identity theft. These services seek a more sinister goal in hiding traffic from law enforcement agencies. Like so many other things, it is this menacing intent that creates the problem and not the site itself.

† † † † †

RANDOM TECH TIP:

Not necessarily TECHNICAL, here is a little hint I picked up that I wanted to pass on to each of you. Have you ever bought some nice and nifty cool "goody" only to practically slice off a finger or two trying to get into the plastic clamshell package? Next time you are faced with this situation, reach into the kitchen drawer and try out the good old manual can opener. It works wonders on most of these packages.

I said Internet monitoring is one of the most hated systems running in my office. Why is that true? Besides it inconveniencing Fantasy Football players during the NFL season, since we block game sites, it also can block legitimate sites needed for work. While to you this may be a major deal, and seem like the end of the world, it has actually become quite rare as the classification systems have become more accurate. Good communication can help minimize the trauma!

There is also another important instance where the blocking of a legitimate site can be to your advantage. Besides just blocking sites based on categories in a definition file, Internet-monitoring software can also block a site based on suspected malicious activity. The unfortunate truth is that legitimate sites are compromised. A site that yesterday was safe can become a nasty source of Malware today. Your computer may be trying to help you out!

We can see how Internet monitoring and blocking at work can help prevent malicious and non-work-related information from reaching and displaying on your computer. How about at home? Are we stuck exposed to all the Internet can throw at us? Luckily, the answer is no, we are not. You can provide your home network the same protections you receive in your office.

There are plenty of services available to home users that will help you accomplish protecting your family and home network from content you object to on the Internet. It is important to make THAT distinction: content YOU object to displaying. My rules and your rules for your home may be different. For that reason, I will not go into detail dictating what categories you should prohibit. That decision is yours. Any solution should allow you that flexibility, in my honest opinion. What I WILL disclose here is the tool I have chosen to use to protect my family and friends.

For my home, I have chosen to utilize the services of OpenDNS®. They have a couple of options available, of which the core services for a home user are FREE! This will provide you with the ability to decide the level of protection you desire along with any exceptions you feel are necessary. Even the paid version is less than $20.00 a YEAR, which will allow you more reporting and exceptions for your filtering rules. For those interested they also provide corporate services for additional fees.

To understand OpenDNS® we must first think about what DNS does within our network, as discussed in the home-networking chapter. The DNS service takes the name you type into the address bar and converts it to a number the computer can understand. OpenDNS® inserts itself into the process by evaluating the sites requested and blocks based on YOUR preferences.

The OpenDNS® web site (***www.OPENDNS.com***) has excellent tutorials on the simple process of setting up their service for use in your home. There are visual guides for making the needed simple changes to your specific operating system or networking equipment. Their site can guide you much better than I could here in a few pages. Once you complete your setup you can select from a list of approximately 30 categories to begin immediately taking control of the web sites permitted in your home.

It is important to remind you of a very important computer security fact. OpenDNS® does NOT replace the need for other protections such as Anti-Malware software. It simply becomes another layer in your protection. A common analogy when discussing computer security is like the layers of an onion. By applying multiple layers, you increase your strength and effectiveness.

My Computer Has Cookies?

What do browsing the Internet and the Keebler® Elves have in common? Both deal with cookies. However, the cookies in your computer would not do well if you left them out for Santa. Computer cookies are small pieces of data left behind by a web site to record your visit and remember any preferences you selected when visiting the site. For example, if you were visiting a web site with multiple language options the cookie would record your selection.

These little chocolate chips are not all safe. Programs also use cookies to track your activities and habits. Their presence on your computer leave a "*trail*" to follow, which gives rise to privacy concerns. The "*Tracking Cookie*" is the primary culprit in this particular activity.

Another "*flavor*" of cookies on your computer is the "*Authentication Cookie*." These store your logon information, such as username and password, to permit rapid login to sites you frequently visit. These cookies tend to crumble when intercepted by a hacker, leading to the compromise of a user's information. More cookies exist, such as session cookies, persistent cookies, and even SUPER cookies. No need to bore you with the details. What we are most concerned with learning is the successful management of cookies to protect our computers and ourselves.

While they serve some useful purposes, it can be in our best interest to clean out the cookie jar from time to time. While there are plenty of tools out there to tackle the task, I have a favorite program I use when I am looking to clear out cookies and other temporary files from my computer. A wonderful company named Piriform (**www.PIRIFORM.com**) calls it "*CCleaner*".

CCleaner is a popular tool used to clear many different temporary files from your computer and help to protect your privacy. I have found it safe and simple to use. The basic version is FREE (*Moving up to the paid version (at the time of this writing, $24.95) will add automated updating and support options*).

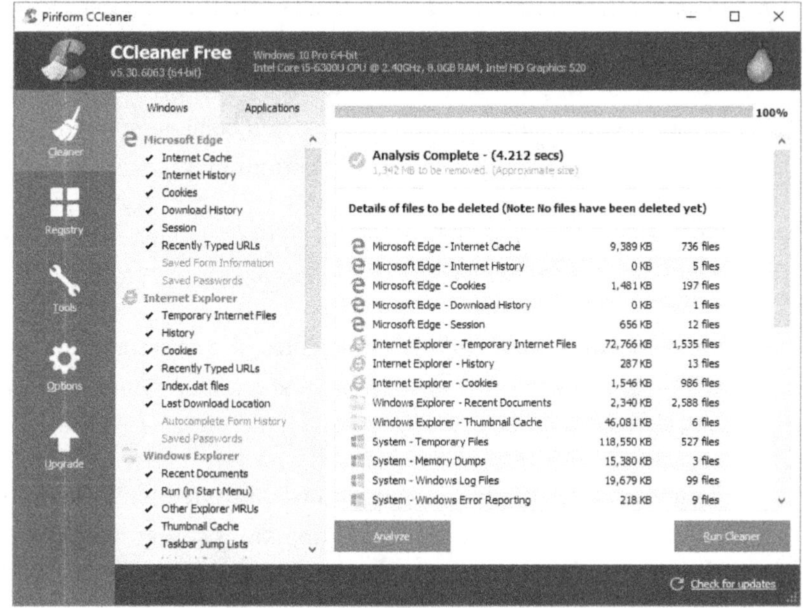

Sample Screenshot from CCleaner

As you will see, there are quite a few temporary files CCleaner will safely remove from your computer. The recycle bin, program temporary file, and other unnecessary files can be cleared with the click of a button. It goes well beyond just the cookies! I run this program once a week as a part of my regular maintenance routine, finding the free version to be more than adequate for my needs.

Secure Web Browsing

Have you had people tell you to make sure a site is "*secure*" before putting in your information, especially when dealing with a bank? How can we tell if the site is "*secure*" or not? There are a few visual indicators to help. Some examples:

- A padlock indicates to the web browser that the site is secured with an SSL (*Secure Socket Layer*) certificate, a computer version of a certificate of authenticity. Depending on the browser you use the padlock image could be an icon on top near the web address itself or along the bottom in a corner.

- The web address will begin with HTTPS instead of HTTP. The added "S" indicates the site is "*Secure*."

- On some sites, especially financial sites, the entire address area will turn a shade of green. This is yet another visual indicator of a secure site.

Your computer WANTS to tell you when something is not right. It is up to you to pay attention to the warning signs if you want to stay safe. One of the common warnings I have seen numerous people simply click through occurs when a computer checks the authenticity of the sites certificate, realizes something is not right, and warns them. The user then "*assumes*" it must simply be an oversight and continues against the warnings.

∞ ∞ ∞ ∞ ∞

"Home computers are being called upon to perform many new functions, including the consumption of homework formerly eaten by the dog."

– Doug Larson

∞ ∞ ∞ ∞ ∞

There is a problem with this website's security certificate.

The security certificate presented by this website has expired or is not yet valid.

Security certificate problems may indicate an attempt to fool you or intercept any data you send to the server.

We recommend that you close this webpage and do not continue to this website.

Click here to close this webpage.

Continue to this website (not recommended).

More information

Sample Internet Explorer Certificate Warning

Another common issue facing Internet users today is the ease in which a malicious person can create and promote a fake site to lure victims. To prevent arriving at a site that is not legitimate, you should never follow links to sites contained in email messages. It is simple to fake links and take you to a forged site. This is especially true with financial sites. It is always best to manually type the address yourself to ensure you end up on the proper site.

Perhaps the best advice to follow when browsing the Internet is to pay attention! If something does not seem right, it likely is not right! If you were to arrive at your home or car and got that same feeling, you would take precautions. We need to give that same respect to Internet browsing. Unfortunately, there is a criminal element present and more than willing to take advantage of the naive. Legitimate businesses and web sites that want to earn and maintain your business will take measures to insure your safety. If this is not the case, I would suggest they may not be worthy of your patronage.

Toolbars

Toolbars are those little strips of buttons and boxes you sometimes see on your Internet browsing program screen. Their appearance gives computer geeks nightmares. So many times when we hear someone complain about a slow computer, we find these little demons at the root of the problem. These add-on items come packaged with some software to "*enhance*" your Internet experience.

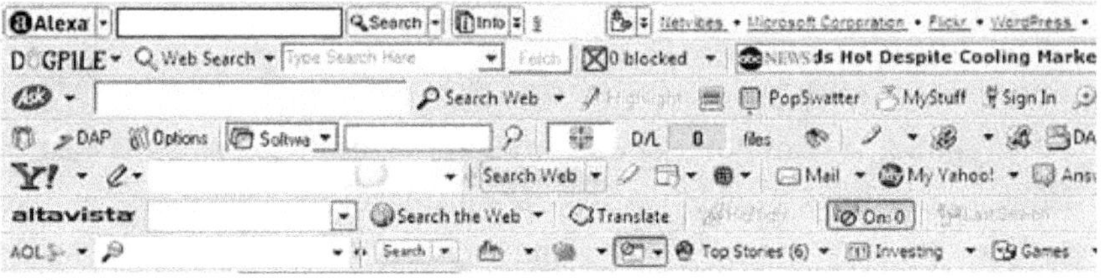

Browser Sample Overloaded with Toolbars

While at first glance it may seem these programs are helpful, the reality is they are gossipy. They can steal performance from your computer and Internet browser. They do such things as monitor your browsing habits, reporting sites you visit back to manufacturers in order to send you more targeted ads and pop-ups. In my opinion, they are at the minimum "*spyware*" and I would recommend avoiding them when installing any software. If you find they "*appear*" on your Internet browser, I would suggest removing them as soon as possible.

Technology Care and Feeding

"A computer once beat me at chess,
but it was no match for me at kick boxing."
– Emo Philips

Thankfully, your computer does not require an oil change. Nevertheless, it does need your attention. Unfortunately, many consider their technology items, especially their computers, as a "*Set It and Forget It*" style of device. Nothing could be farther from the truth. As we have seen throughout these chapters, there are many simple and FREE things you can do to take care of your investment. After all, it IS an investment. I doubt any of us can simply afford to run out and buy a new computer each time our current one hits a snag. In this section, we will cover some maintenance topics to help round out your maintenance needs with your technology.

Side Note*: If anyone reading this CAN afford to throw money away, please contact me regarding my "Money-Recycling" program in which you can deposit any spare money cluttering your home or office!* ☺

A common question is the continuation of the debate of whether or not it is better to turn off our computers or leave them running. It is the "*Paper vs. Plastic*" of the technology world. The truth is there is really no "*right*" answer. There are valid points on each side of the debate and are best decided by you and your usage patterns and requirements. Some of those points are:

- Power savings when turning off
- Less wear and tear when leaving on
- Less hacker access by turning off
- 24-hour access when leaving on
- Alleviate heat issues by turning off

While these are valid points in some respects, I tend to have my own opinion, which OF COURSE, I will share here with you, my dear reader. Your opinion, as well as your policy at work, may vary. For my home, as well as in my office, I prefer turning off systems EXCEPT for cases where after-hours access is required. Understandably, we have systems in the office that need to be accessed 24-hours a day, such as web servers. However, general users do not typically require such access with their desktop computers.

For the home user you actually have options available that can give you the best of both worlds. Many computer systems can be set to turn themselves on at a pre-determined time. For example, I have some friends that work from home and heavily rely on their computer. They turn their systems off each night, but prefer waking up, stepping into their office, and having their computers on and waiting for them. To accomplish this task, we simply set the computer to have a wake-up time. How you set this on your particular system varies between manufacturers. In my friend's systems we made the settings in the computer BIOS, which is the basic system built-in to the hardware of every computer. Not ALL computers support this feature, though many newer models do. Consult your manufacturer's documentation for specifics on your system.

Another option available through the operating system for most modern computers is the "*Hibernation*" and "*Sleep*" mode settings. Though these two options vary slightly, their principles are the same. In both, your computer descends into a low power setting, conserving energy.

In "*Hibernation*" mode, the power to your computer completely shuts off, as if you turned it off. The computer takes all currently running programs and other open objects and stores them in a temporary file on your computer. It then shuts down your computer. Once you "*wake*" your computer it restores everything as it was previously and you continue where you left off. Even documents and programs you were actively working with will be restored exactly as you left them.

While in "*Sleep*" mode, your computer goes into a reduced power state similar to the pausing of your DVD player. Once you press a button to "*Wake Up*" your computer instantly turns everything on and resumes right where you left off. This option is much faster to restore and resume working than hibernation. The tradeoff is that it will still consume a little bit of power.

RANDOM TECH TIP:

Passing along each email you receive to friends and family regarding a free offer for store gift cards does nothing but clog up their email inboxes. Unfortunately, companies DO NOT offer gift cards no matter how many times you forward their message. This is a hoax meant to waste your time. I suggest people check with a reputable "rumor control" web site like ***www.SNOPES.com*** *prior to passing on the latest "too good to be true" or other such offer. This includes notices about missing persons. While occasionally these may be legit, they are oftentimes either outdated or downright fake.*

Aging Technology

Sometimes your performance issues can simply be because your computer is getting too old. While I firmly believe we can take some actions to extend the life of technology (*such as increasing the system memory, which we will discuss on the next page*), we sometimes simply must realize when it is time to throw in the towel and move up to a newer model.

Over time, it becomes harder to get and support older options. It is just not economically feasible for the technology vendors, such as printer manufacturers, to continue to develop for "*old*" technology. There is an expectation in the market that users will keep up. The sad truth is there is just not a market for "*vintage*" computers.

The fact is that technology moves at a lightning pace in our society. There is always a new model coming out. We see new features make it onto the showroom floor and suddenly last year's big deal becomes today's antique. Even the best of the best computer nerds know there comes a time when an update just will not make a difference. Replacement should never be your first option, but it does remain a potential contender.

Upgrading Memory

Earlier we compared your computer to a cluttered desktop. We mentioned how the rebooting or restarting of your computer would clear off that desktop and give you a fresh area to work. The "*desktop*" area of your computer we were tending to is the RAM, or memory. This is a component of your computer that tends to be easy to upgrade and can add years of life to your equipment.

Imagine if you were trying to do all of your work on a single TV tray in your office. Your space for working would be significantly small. You would be crowded and would have difficulty finding space to work. This is how your computer "*feels*" when it has inadequate memory. Luckily, it is easy to move to a bigger desk.

Upgrading the RAM (*memory*) in your computer is one of the simplest and cost-effective steps you can take. I always suggest people consider this option before moving on to buying new equipment. It is even easy to find out EXACTLY what you need without having to be a computer engineer. If you guessed that I was going to give you a recommendation how to do this, you have guessed correctly!

I personally use a company called Crucial® (**www.CRUCIAL.com**) when looking to upgrade memory in a computer. This goes for desktops, laptops, and servers. One truly nice feature of their web site is a small program to run on your computer to help evaluate exactly what you need to be 100% compatible with your equipment. It is their "*System Scanner.*" It will evaluate your system, tell you what your options are, and give you suggestions for upgrading your computer. It will also remove the scanning component from your computer, thereby not leaving yet another program cluttering your computer. As Crucial® is also a manufacturer you can then buy the option best suited to your needs and budget.

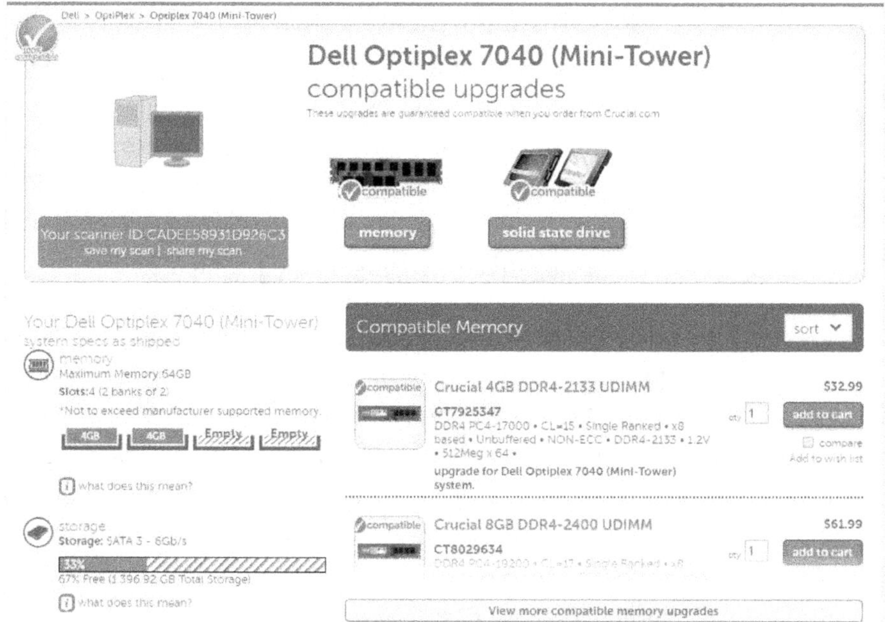

Screenshot of Crucial System Scanner Results

Once you order your upgrade, installation is generally easy. Most computers have an easily accessible area (*even laptops*) in which you can install your upgrade. On the web site, there are video tutorials to assist you in self-installing your memory upgrade.

† † † † †

IMPORTANT NOTE:

Many new computers, especially the ultra-thin tablet-like units, have their memory permanently attached to the main board in the unit. This permanently attached memory IS NOT user replaceable or upgradable. The scanner will tell you if this is the case with your device, typically by saying it is an unsupported system for a memory upgrade.

Dealing with the Clutter

Like a neglected closet, we end up with piles and piles of things installed on our computers. It could be games or programs used long ago, which have been forgotten. It could be add-on programs that came along as we installed another program. However, it happened, the fact remains that over time our computers get cluttered. Just as we dread the cleaning of a closet, many dread cleaning up their computer.

This process is much easier with a few simple tools. The basic features for removing a program are built into most operating systems (*the feature is found inside the system "Control Panel" named either "Add/Remove Programs" or "Programs and Features", depending on your version*). To simplify the process, and instructions, I am going to direct you to a program we previously discussed named "*CCleaner*." You should recall this program from the web-browsing chapter when we discussed cleaning temporary files from your computer.

CCleaner has additional functionality beyond simply cleaning temporary files. One of those is the easy to find and use "*Uninstall*" function. You find it by clicking on "*Tools*" in the left column of the program. Once there you can simply scroll through the list of programs installed on your computer and select any you wish to remove. Clicking the "*Run Uninstaller*" button will start the process of removing the unwelcome program.

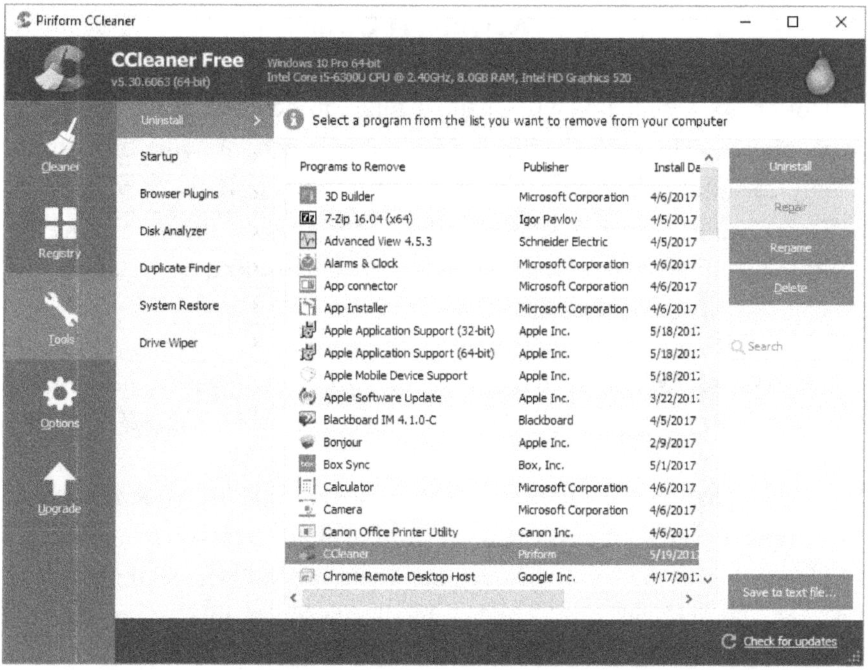

Screenshot of CCleaner Uninstall Section

RANDOM TECH TIP:

Just because an email appears to be from a person you know and trust, still use caution when opening or following instructions in the message. Malicious software can compromise a user's email address book and pretend to be anyone of the addresses it discovers. This means that if a friend or business associate has his or her computer infected, the malicious software can pretend to be anyone in his or her address book. Bottom line: Even if the message appears to be from someone you know, please proceed with caution.

Keeping It Clean

Dust and heat can be detrimental to electronic equipment. I often joke with my wife that dusting can actually be BAD for her computer. Thus far, she has not let that stop her! As dust floats in the air, the assorted fans running to keep our stuff cool suck it in. Dust, being a great insulator, then contributes to things heating up. If you notice your computer suddenly sounds like a jet taking off, you may have dust issues.

WARNING: *Before following the tips below PLEASE be sure to TURN OFF your equipment to prevent damage!*

One thing you can do to remove some of your dust from computers and laptops is use the little hose attachments on many vacuum cleaners. Simply running over the vents around your equipment can remove quite a bit of the coating dust leaves on surfaces. You can also use the blower feature on some vacuums to give a quick burst of air to dislodge the stubborn dust. I have even seen someone take the computer outside and use the leaf blower! A little over-the-top, but I admire their style!

There comes a time when you have to get more aggressive with a dust problem. This may involve the opening of the equipment to get inside and clean things up. **IMPORTANT:** *I would NEVER recommend a regular user open up their laptop. Even those with experience in such things cringe at the thought of doing such a task.* Your desktop computer, however, tends to be an easier beast to tackle. After making sure the computer is off, gently use your vacuum to clean inside the box. Some people also use cans of compressed air for those hard to reach nooks and crannies.

Beware of the Bundle

When installing software, you should pay attention to what else the program may try to install. Many programs today come "*bundled*" with additional products that may cause issues in performance. At the very least, they could be programs you are not interested in using and would take up space on your computer. In most instances, it is simply a matter of clearing the check box to prevent the installation of this extra software. Here are a few more tips for avoiding the bundle:

- Choose "*Custom*" install if available, as many times the bundled software option is "*hidden*" under this option.
- READ each screen prior to clicking "*Next*" or "*Agree*". As I tell my kids…simply pay attention.
- I know most people pass by it, but it is in YOUR best interest to READ that long list (*commonly called the End User Licensing Agreement, or EULA*) to see what you are agreeing to before clicking.

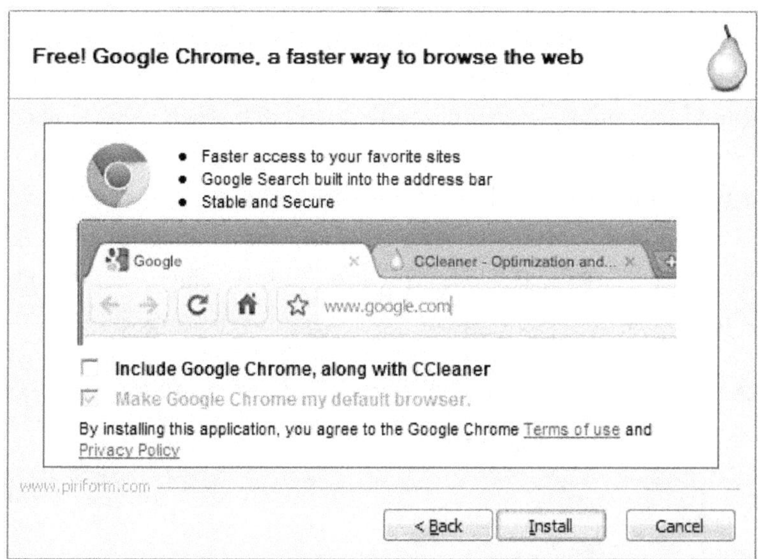

Sample Image of Screen Attempting To Install Add-On Software

Backing Up Your Stuff

At the beginning of each class I teach, I ask students one basic question: Do you backup your computer? It never ceases to amaze me when barely a hand rises into the air. This phenomenon is even more surprising when the classroom contains "*nerd*" students seeking education in computer fields. Crazy, but true.

Many people incorrectly perceive backups as a difficult process for the home user. This could not be farther from the truth. There are many easy options available to provide you with total protection in case of a system failure. The sobering reality is if you have not experienced a failure of your computer, you are in the minority. They do break, and when they do, you run a high risk of losing everything. The wrong time to think about a backup is AFTER you have experienced a loss. It would be as if trying to purchase car insurance after a texting teenager hit you, with similar results!

As you have by now come to expect, I have a recommendation for this subject as well. Actually, I have two on this topic. With backups, I prefer not to keep all my proverbial eggs in one basket. The primary and most important items on my home computer are the multitude of digital pictures my wife and I have taken of our kids since they were born. While I am not sure what the penalty would be to THIS nerd if anything were to happen to those pictures, I do know it would likely result in my sleeping with one eye open and dodging a frying pan to the skull! (*Love you, Honey*!) I use the following two methods for backing up my home computer:

- External drive physically connected to the computer
- An online service to provide "*off-site*" storage

My primary option is my external drive. These have become much cheaper over the years, providing a backup solution I can physically touch. Each manufacturer develops and includes their own backup software that is easy to use for even novice users. Some of the advantages (*and disadvantages*) of the physically attached external drive for backing up your computer are:

- Maintain physical control
- Operation not dependent on Internet connection
- Vulnerable to natural disaster or theft

As a further line of protection from the previously mentioned application of a frying pan against my thick skull, I also employ an online backup solution. My personal choice is Carbonite® (**www.CARBONITE.com**). I chose them due to their reputation, reliability, ease of use, and affordability. At the time of this writing, the cost per year was $59.00 for unlimited storage space for one computer. A small program installs on your computer, which ensures your data stays synchronized with the online service. Some of the advantages (*and disadvantages*) of online backups are:

- Physical separation of your data in case of disaster
- Online vendor responsible for storage
- Dependent on Internet connection

While either of these solutions by themselves is adequate and better than no action at all, I feel the combination of them gives me a far greater level of protection. We see data is growing day by day, so the protection of that data will continue to grow in importance.

Embrace Your Local Nerd

Finally, yet importantly, I wanted to throw in a plug for your local computer nerd. While it can be convenient to head into a big store with your technology woes, please consider that person with the small shop on the corner or in their garage. Computer nerds and geeks have a long, distinguished history of being self-starters and innovative. In fact, many of the "*large*" companies today grew from such humble beginnings. In my experience, I have come across many with a sharp mind, helpful attitude, and true desire to help. Doing a good job for you means so much to them, as you become a powerful advertising force. Talk to friends and neighbors and see what their experience has been, and do not be afraid to share your own when you find that gem of a technician that takes care of you.

RANDOM TECH TIP:

Are Screensavers still important?

*I hear this question quite a bit. Back in the days of the CRT monitor (*the ones that looked like big, fat TV's*) the problem existed where an image could be "burned" into the screen. Some of you may remember seeing distorted, "burned" images on older ATM machines.*

*Current LCD (*flat panel*) monitors do not have this same problem, making the screensaver not as important.*

HOWEVER...don't be so quick to turn it off!

Screensavers still serve a valuable purpose beyond showing a slideshow of the kids or grandkids. You can set the screensaver to lock after a period of time, forcing a password to resume using your computer. This helps prevent people from sneaking onto, and potentially abusing, your computer while you are away!

Frequently Asked Questions

"One of the most feared expressions in modern times is 'The Computer Is Down.'"
– Norman Ralph Augustine

I threw a question out to friends and family on Facebook© asking what was their biggest frustration with their computers. I thought answering some of those issues from my experience might help you as well! These answers are according to MY opinions and experiences. I hope they will help clear more of the fog surrounding your computers and why they do the things they do.

Q. Why does my hard drive (or thumb drive) show one size/capacity on the package though a different, smaller size/capacity when I put it in my computer?

A. The root of this problem is MATH. Computer operations are based on binary, or "*Base 2*" (*only uses the numbers 0 and 1*) while we typically think in "*Base 10*" (*using numbers 0-9*). Therefore, what we see is a "*difference in communication*" which results in the numbers not being evaluated equally. Moreover, because of the difference in the math foundations you cannot simply count to 1,000 to equal the next step up. For example, a kilobyte is ACTUALLY 1,024 BYTES (*A BYTE is a computer unit of measure to describe the amount of space data consumes*). Using the same numbers, we can continue "*up the ladder*", so to speak:

- Kilobyte (KB) = 1,024 Bytes
- Megabyte (MB) = 1,024 Kilobytes or 1,048,576 Bytes
- Gigabyte (GB) = 1,024 Megabytes or 1,073,741,824 Bytes
- Terabyte (TB) = 1,024 Gigabytes or 1,099,511,627,776 Bytes

Therefore, what you see on the box when you buy an item is that the manufacturer MARKETS the device with the higher number, whereas the computer recognizes it based on its own digital calculations. For every gigabyte reported on the box, your computer will recognize approximately 70 megabytes less in capacity.

Q. What do error messages, like "*Fatal Exception Error*", mean?

A. Let's first say this does NOT mean, necessarily, that your computer is about to die a slow and agonizing death. A Fatal Exception Error is a warning from a program that is has to stop because a mistake has occurred. This relates to the program itself and normally indicates a programming error. This is TYPICALLY not a cause for major alarm, despite the lethal wording.

The more troublesome error message is the dreaded "*Blue Screen of Death*", where the screen goes blue, a bunch of text that likely makes no sense appears, and your computer shuts down or restarts. Often this is an indicator of eminent hardware failure, although it could be an issue with the operating system. I would recommend reaching out to a local, reputable techie to help diagnose these issues if they become habitual. A "*Once in a BLUE Moon*" (*pun intended*) occurrence is typically nothing worth being too worked up over. Think of it like a cough: Not every cough requires a visit to the doctor, but when it gets worse or more persistent, it is time to check things out further.

† † † † †

RANDOM TECH TIP:

Remember quite a few pages back when we mentioned the spiders web of wires we find hiding around the desk? Besides aggravating my wife with their unsightly appearance, they also provide plenty of aggravation when trying to figure out what goes where. Using the little plastic tabs that come on your bread loafs can give you a quick and easy way to label each cable. A permanent marker can label things such as "Printer" and "Keyboard", making things so much easier when digging your way through the web.

Q. Is public Wi-Fi safe?

A. Let's answer the question with a question: is swimming in the ocean safe? Sure it is, until you come across a shark that has not eaten in a couple days and thinks you look a lot like a buffet in a brightly colored swimsuit.

Much like the ocean there is a constant threat that surrounds you. However, unlike the ocean, this threat is more widespread. While your chances of being attacked by a hungry shark are one in a million, your chances of being attacked on public Wi-Fi are much, much greater. Once you place yourself on a public network connection, you are truly opening yourself up to the potential for infections and compromise, as anyone else with a wireless-capable device can potentially "*see*" you. Am I therefore saying NOT to use publically available access, such as in restaurants, airports, and coffee shops? No. Going back to our ocean example, you would not avoid the ocean all together because of the potential dangers, but you also would not jump in wearing a seal-shaped suit while covered in fish bait in the middle of a feeding frenzy. (*OK, maybe one or two of you would, but...)*

† † † † †

RANDOM TECH TIP:

If you receive a phone call or email from your bank, credit card company, dry cleaners, or mother's cousin's nephew's former roommate stating they need your password to tend to an issue, correct a problem, or simply verify your account, it is a LIE! Legitimate companies with which you do business do NOT need your passwords. They already have your account information, which is all they need. Protect your passwords or PIN numbers as you would your social security number.

It would be in your best interest to be AWARE of the dangers and take a few precautions:

- Do not perform financial transactions. With the risk of exposing information, the threat to your fiscal identity is great.
- Make sure all of your software, especially security software, is up to date.
- When possible, use a VPN (*Virtual Private Network*) connection to connect to a trusted network, such as at home, to conduct your work more securely.
- When you connect to Wi-Fi, a box similar to the one shown will appear asking you to identify which type of network you are connecting to at that time. Choosing "*Public*" will auto-magically make certain security settings on your computer to prevent malicious users from easily attacking you.

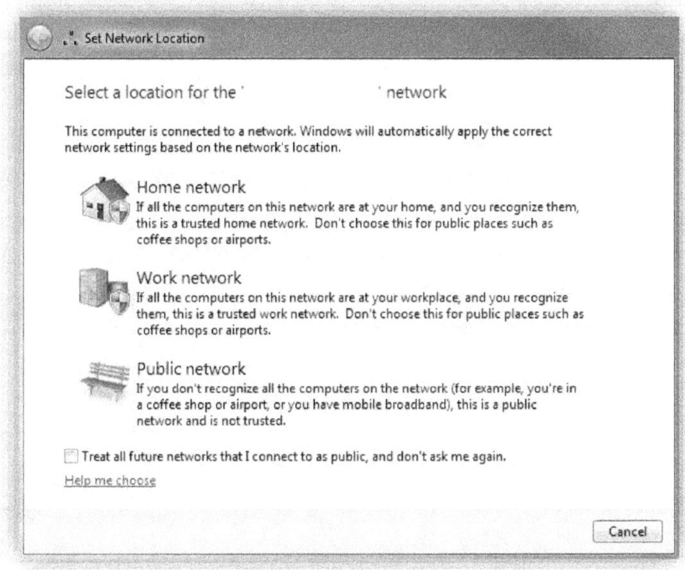

Sample Network Location Setting Pop-Up

- If you receive pop-ups or warnings advising you of CERTIFICATE ERRORS on a web page, CLOSE the page. Often this can be an attempt by a malicious person to compromise your system. Legitimate sites should NOT present these alerts; this is your computer telling you something is WRONG (*this applies anytime you are surfing the web, including from home*).

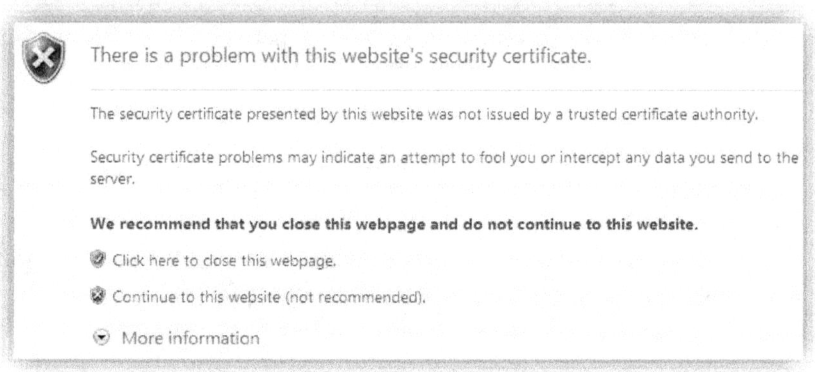

Sample Security Certificate Warning Message

I only utilize public Wi-Fi sessions, personally, for random web surfing, such as for directions, reading reviews, or checking movie times. I prefer to only enter my information (*passwords, etc.*) when I am on networks I "*know*" and can "*trust*" with relative certainty, such as at my home or office.

Q. Can these commercials that offer to increase my computer speed really do what they say?

A. Short answer: NO! When I see these infomercials that promise to double your speed or make all your computer issues go away, and frankly, it infuriates me. It does damage to the reputations of the honest techies in the market trying to build a reputable business. These online services install their own version of software to give you a false sense of security, when in reality they may be causing MORE slowdowns, all while possibly spying on you. There is NO magic bullet to maintaining your computer, and in MY opinion, these infomercial companies will cause more harm than good. I have presented several reputable, and often free, tools in this book to help YOU take control of your computer. Following these guidelines, along with engaging reputable technicians for advanced help, will save you time, money, and frustration.

RANDOM TECH TIP:

*Did you delete an item and realize you NEED it back? Most of the time you can go into your "Recycle Bin" and recover the lost item. However, sometimes this is not the case. In situations like this I recommend people try another free tool from the creators of CCleaner. At their same site (**www.PIRIFORM.com**) you can get a wonderful program named "Recuva". It is a simple to use program that will dig in and find many deleted items. It is not a replacement for being careful, but if used quick enough after an accidental deletion it can save the day.*

Q. What is the thing that pops up when I try to run a program that asks for my permission?

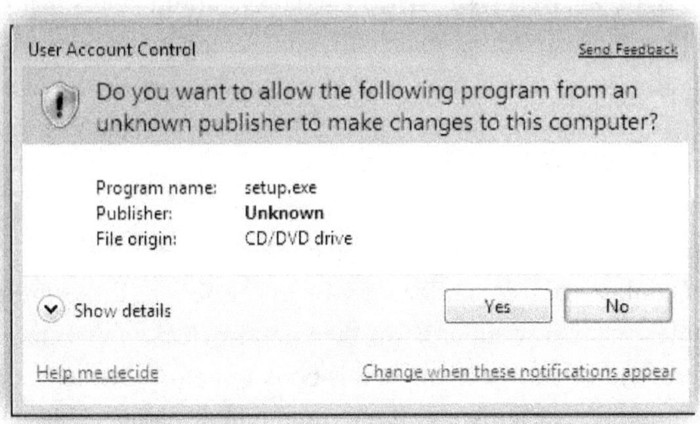

Sample of a User Account Control (UAC) Pop-Up

A. The program itself is the "*User Account Control*" or UAC for short. Simply put, it is a Windows® feature meant to prevent unauthorized changes to Windows®. I think of it as a "*Mother May I*" sort of thing. The idea behind it is that by taking the extra step and verifying you REALLY meant to run the program it will prevent AUTOMATED programs from being able to make changes, or installing malicious software, to your computer.

Is it effective? It does a good job, though it is not perfect. Again, it is not a cure-all, but it can help in your fight against infections from unwanted software. I see many people choose the option link in the lower-right corner of the pop-up to change when the notifications appear, typically disabling the protection. While I agree UAC can get aggravating at times, I believe the extra protection is worth the effort.

Q. I hear your fellow nerds mention something about "*drivers*" and the issues they cause. What do they mean?

A. Drivers, in their simplest form, are programs that allow your operating system, such as Microsoft® Windows®, to talk to your hardware components, such as your printer or video card. For example, when you bring home a new printer you would run the CD that came with it. On the CD are the programs that will give you the advanced features. The most important thing on the CD is the small program that tells your computer how to talk to the printer in the first place. This describes the BASIC principles on how they are to communicate. If this program is not installed, the computer will fail to recognize the device.

Like other programs in your computer, drivers occasionally need updating. Many may appear as a "*Recommended Download*" during the Windows® Update process. These are generally safe to install, though I generally prefer seeking them from the manufacturer. Your manufacturer's web site will contain the drivers for your specific model of equipment as it becomes available. I would consider this the best source for quality-tested drivers.

NOTE: *Never trust drivers resulting from an Internet search. Many of the located web sites are malicious Always use either the manufacturer or Microsoft® for safe driver downloads.*

∞ ∞ ∞ ∞ ∞

"Information on the Internet is subject to the same rules and regulations as conversation at a bar"

– Dr. George Lundberg

∞ ∞ ∞ ∞ ∞

Q. I keep getting these pop-ups telling me things need to be updated and such. How can I tell the real ones from the fake ones?

A. This can be a difficult question to answer, as there are so many possibilities. However, the first thing I would advise is to KNOW what you have installed. I have seen many instances where people have infected their computers clicking on a pop-up stating that a program THEY DID NOT EVEN HAVE INSTALLED needed an update. Quite a bit of common issues can be prevented by simply being aware of the programs you ARE running or have installed!

There are, however, many legitimate times your computer WILL alert you when it requires an update. For example, Microsoft® Windows® will generally prompt to install updates and restart your computer around the second Tuesday of each month as they release the updates to their systems. Other programs such as PDF readers will ask for the same thing. This is expected, and you should allow them to update to keep a solid defense against new threats. Knowing WHAT programs are installed will help you determine the relative legitimacy.

One of the most important pop-ups you need to be aware of and familiar with is the one associated with YOUR security program(s) when it wants to alert you to a possible infection. There are MANY different manufacturers and versions of security software out there… excessively numerous to specifically list here. Your best action would be to visit your chosen solutions web site and become familiar with the assorted alerts you may receive. Each manufacturer typically will have a "*Frequently Asked Questions*" section to assist you.

Java Update - Update Available

Java Update Available

Java 7 Update 05 is ready to install. Installing Java 7 Update 05 will uninstall the latest Java 6 from your system. Click the Install button to update Java now. If you wish to update Java later, click the Later button.

Java

More information...

ORACLE

Install Later

Windows Update

Restart your computer to finish installing important updates

Windows can't update important files and services while the system is using them. Make sure to save your files before restarting.

Remind me in: 10 minutes

Restart now Postpone

Adobe Reader Updater

An update for Adobe Reader to version 10.1.3 is available. Do you want to install it now?

This update addresses customer issues and security vulnerabilities. Adobe recommends that you always install the latest updates.

Details

No Yes

Samples of Update Pop-Up Notices

Q. I see where people advise NOT running your computer as an "*Administrator*". What does this mean, and why is it so bad?

A. As an "*Administrator*" on a computer, a user has full and unrestricted access to do as they please with the computer. This means installations of software and changes to system security settings. Therein is the reason it is so dangerous. As the access is unrestricted, the potential for a malicious program or web site to exploit your level of permissions is high. What is the result? An administrative user is more likely to become infected with malicious software, or install something that slows down their computer.

Now, there are times you NEED those administrative rights. When you are installing new software, or updating existing software, these privileges are required. However, for general use there is no consistent need to run as an administrative-level user. With Windows® it has become much simpler in recent years to create a "*standard*" user. When running as a standard user, you will be prompted anytime you NEED the elevated administrative privileges to enter the password for the administrative user. This will help prevent malicious programs from installing themselves in the background and gives YOU control over any installations.

Will this make your computer impenetrable to malicious software getting in and messing things up? Unfortunately, it is not 100%. I can tell you, however, that it WILL make a SUBSTANTIAL positive impact on your computer experience and block almost all the nasty little cooties that try to slip inside. I have seen offices go from multiple infections a week to virtually none at all with this simple change. Many users fail to realize the difference once they have made the appropriate settings which makes this simple change painless.

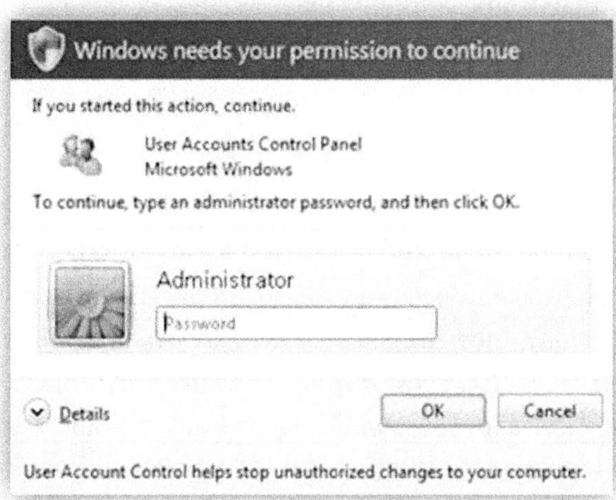

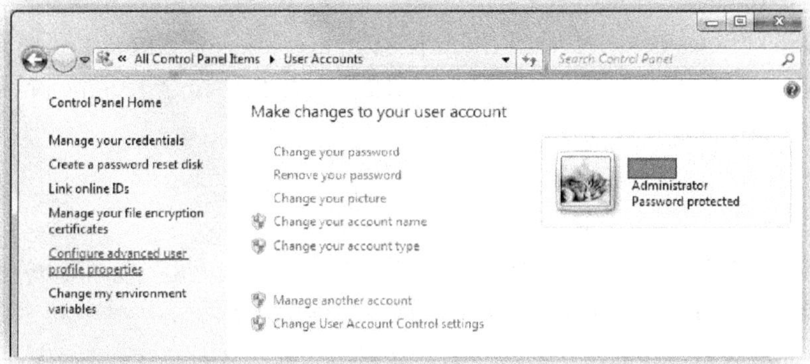

Sample of Prompt to Supply Administrator Password (TOP)
Screenshot of User Accounts Section in Microsoft® Windows® (BOTTOM)

To be effective you need to set a password for your administrator account and then STOP using it on a regular basis! In Microsoft® Windows® it is also easy to set accounts as a regular, or "*Standard*" user, as us techies would call it. We do this in the Control Panel "*User Accounts*" section. In the illustration, you can see the links you can click on to "*Change Your Account Type*" as well as to change, or set, your password.

Q. I feel like my computer tech is always talking down to me. Is there anything I can do?

A. While not a technical question, per se, it is still related. Unfortunately, there are people out there that are not the best at customer service. We see this in SO many industries today, and it is unfortunate. Do not let them mistreat you! There are plenty of others out there that would be PLEASED to help you. Ask around to family and friends for someone they may know who is a QUALITY person. While knowledge is wonderful, it is NOT everything. There is NO excuse for poor customer service, and you should not reward it with your patronage. In addition, I typically have found the know-it-alls really are not THAT good at all! You will get better results with someone who understands you, the user, as well as the technology, and who is willing to work with both. It upsets me as well when I see it, and it is something I hope this book can help prevent. Nothing throws off the smug technician like tossing some knowledge back at them.

BLATANT PERSONAL PLUG: *My technology company, TechKnolutions, can assist you with computer repairs, or even help you locate a local technician. My associates or I will gladly assist you, even remotely. Contact information available on the Copyright page.*

A Few Definitions

"Buying the right computer and getting it to work properly is no more complicated than building a nuclear reactor from wristwatch parts in a darkened room using only your teeth."
– Dave Barry

OK…you can admit it. Your neighborly computer nerd is full of (*do not say it*) …acronyms and strange terms that we do not possibly know the meaning of without an advanced degree in Google-ology. I have assembled quite a few of these here to help explain and make a bit of sense of their meaning.

Access Point – Commonly this is a wireless modem or router placed in your home (*or office*) to allow wireless devices to connect to the Internet; also referred to as the WAP, or Wireless Access Point.

Backdoor – Malicious software can sometimes leave open a "*door*" into your system for the malicious person to access later. Think of it as an alternate way into your system, like the back porch at your home.

Backscatter – Backscatter is seen when you receive email delivery failure notices from addresses you never actually sent an email to in the first place. Besides clogging your inbox, this has the negative side effect of making you THINK you are infected when you are really just the unfortunate victim of SPAM.

Bandwidth – This refers to the amount of speed available to you for Internet usage. As the numbers increase (*along with the price*), things get faster. Think of dial-up as a two-lane dirt road, where as DSL or Cable becomes a six-lane super highway.

Bloatware – The programs or features added to your computer that generally serve no useful purpose. Typically, this appears as trials for programs a company is trying to get you to buy.

Bot – Also called a "*Web Robot*"; a compromised computer under the control of a malicious person(s).

Botnet – Developed when a collection of Internet-connected computers are compromised and controlled by a malicious person(s).

Broadcast – Sending out, such as in the case of wireless router signals. This is the transmitting of wireless signals in the air. Think of these airwaves much like those from a radio station.

Browsing – The act of looking around on the Internet. We refer to this as "*browsing*", just like looking around in a store.

Cable Modem – A form of high-speed Internet provided by your cable company, the modem serves as your connection to the Internet.

Certificates – When dealing with web sites, site owners purchase certificates through trustworthy third party companies that verify the sites ownership by the company. Think of it as a "*Certificate of Authenticity*", as one would issue for sports memorabilia.

Cookies – Small digital files left on your computer while web surfing. Web sites use cookies to remember site preferences, passwords, and other information related to your site visits.

Cybersquatting – Obtaining web sites, usually those formerly owned by legitimate businesses or with names similar to legitimate businesses, often for malicious intent.

Driver – This has nothing to do with DRIVING Miss Daisy. A driver in your computer is a set of instructions that tells a piece of hardware, like your video or sound card, how to "*work*".

DSL – Acronym meaning "*Digital Subscriber Line*". This is the high-speed Internet service provided through your telephone company.

Encrypted – When data is scrambled into unrecognizable "*garble*" for the purposes of hiding the information. It requires a "*key*" to descramble and return to readable form.

False Negative – When a program detects an item as being safe but is actually malicious.

False Positive – When a program detects an item as being malicious but is actually safe.

Firewall – Think of this as a device that sits between you and the gangs of malicious folks on the Internet. It helps prevent unwanted malicious activity from entering your systems. These can be a physical (*hardware*) item, or as software running on your system.

Hardware – Items you can physically touch. This includes things such as your keyboard, mouse, printer, monitor, and the computer itself, including the physical components inside.

Heuristics – This refers to the ability of security software to analyze patterns of a program to determine if it is malicious.

Hoax – A prank or fake message generally designed to waste your time and clog up your email inbox.

Host – We nerds refer to a "*host*" as any item plugged into or used on the network, either wired or wireless. Typically, this is "*hardware*".

Internet Service Provider (ISP) – This refers to the company that provides you with Internet service, typically for a monthly fee. Commonly, this will be with your cable or telephone company but can also be with a satellite provider.

IT – Have you been wondering WHY the IT in the title is in ALL CAPS? Nope, it was not a typo. IT is a common acronym for "*Information Technology*."

Key Logger – Process by which a program records each character typed on a keyboard covertly, for discovering confidential information.

LAN – "*Local Area Network*". For example, your LAN would be all the items connected together and sharing information inside your home network. This includes computers, laptops, smartphones, etc.

Malware – Shortened term for Malicious Software. This is software designed to disrupt your computing experience.

Modem – A device used to connect you to the Internet. Typically used by cable and phone companies, but also used by satellite providers.

Network – Refers to devices connected in your home or office used for the sharing of information. This could be as simple as sharing an Internet connection, or for the sharing of files and programs.

Operating System – Most commonly, this is Microsoft® Windows®, but can be others such as Linux or Apple® Mac. This is the overall program that runs on your computer to give primary functionality. It interacts with your hardware and other installed software to browse the web, read email, look at pictures, play games, etc.

P2P – Stands for the term "*Peer to Peer*"; a common method of sharing files among unknown and untrusted users across the Internet. A common use is for the sharing of illegal files such as music.

Passphrase – Sequence of words or other text used in place of a password which is longer, easier to remember, and more secure.

Password – A word or phrase used to protect access to computer resources such as web sites and email accounts.

Payloads – In the world of malicious software, the payload refers to what the program does, such as crash the computer or steal a password.

Pharming – This is the directing of users to a fake web site, which is imitating a real site. This is especially common, and dangerous, with financial web sites, such as for your bank.

Phishing – Refers to fraudulent emails sent appearing to be from legitimate businesses attempting to trick you into supplying personal information. Typically, these attempts are sent in bulk in the hopes of landing a few "*phish*".

RAM – Random Access Memory. This is the temporary area in your computer used to work with files and programs you are currently using. Also referred to as the computers "*memory*".

Ransomware – Malicious software that disrupts your computer and demands you pay a fee before restoring services.

Real-Time – When referring to security software, this is the checking of the file or folder prior to use, or as you use it.

Router – This device connects your LAN (*see definition above*) to the Internet. This device changes "*styles*" between a home user and office user, but the function is the same. It is the doorway between your system(s) and the Internet. In "*Nerd Speak*", we call this a Gateway.

Scareware – Software designed to scare you into taking an action. Normally this is an attempt to get you to buy software that you do not need. It is also often itself malicious.

Search Engine – This is a web site used to help you search the Internet for information you are looking for. One of the more popular search engines is found at **www.GOOGLE.com**, though there are others.

Service Provider – Whom you pay the bill to each month to provide you with Internet services.

Shareware – Someone has developed this software for the explicit purpose of sharing with others. It is free, though some developers will accept donations if you like what they have created.

Software – You install these assorted programs on a computer. It includes items such as word processors, web browsers, games, etc.

SPAM – Unsolicited email. You would have to be living under a rock (*no offense to any cave dwellers reading this now*) to not have received SPAM in your email. It seems to clutter many an inbox, and in the case of my mother-in-law's email, never deleted! ☺

Spear Phishing – An extension of techniques used in Phishing. Spear Phishing targets an individual or company specifically with fraudulent emails sent appearing to be from legitimate businesses attempting to trick you into supplying personal or confidential information.

Spoof – Commonly associated with email, spoofing is the act of pretending to be someone else, such as when malicious software sends email pretending to be you, even though it is not actually you sending the message.

Spyware – Programs installed with or without your knowledge to gather usage information from your computer and report to a third party, typically for the purpose of sending you SPAM or pop-up advertisements as you surf the web or use your computer.

SSID – "*Service Set Identifier*". You assign this "*name*" to your wireless network. It identifies it as yours and makes it easier to find when you are connecting your device(s).

Synchronized – Making sure that data in two locations, such as when backing up, remains a true duplicate of one another.

Toolbar – Narrow bars filled with buttons and boxes added to your Internet browser to "*enhance*" your Internet experience. Many generally consider toolbars a nuisance and a threat to computer performance and privacy.

Troubleshooting – The process of tracing and correcting issues which are causing your computers or other technology items from operating optimally, or operating at all!

User Account Control (UAC) – A Windows® feature meant to prevent unauthorized changes to Windows®. I think of it as a "*Mother May I*" sort of thing. The idea behind it is that by taking the extra step and verifying you REALLY meant to run the program it will prevent AUTOMATED programs from making changes to your computer.

Virtual Private Network (VPN) – This is the establishing of a PRIVATE and ENCRYPTED method of communication across the Internet. A VPN protects your computer communications, keeping outsiders from being able to read your information.

Virus – A program or piece of code that is loaded onto your computer without your knowledge and runs against your wishes.

Wide Area Network (WAN) – Related to a LAN. Whereas a LAN is concerned with the network in your home or office, a WAN refers to the connection of all of our LAN's together to form a WAN...which essentially results in...the Internet!

Worm – A standalone malicious computer program that replicates itself in order to spread to other computers.

∞ ∞ ∞ ∞ ∞

"Those parts of the system that you can hit with a hammer are called hardware; those program instructions that you can only curse at are called software."

– Unknown

∞ ∞ ∞ ∞ ∞

About The Author

As Chief Technology Officer (CTO) and owner of TechKnolutions, LLC, Wiltz Cutrer has more than 20 years in the information technology (IT) field. TechKnolutions, LLC provides networking, security auditing, system repair, and technology consulting along with Continuing Professional Education (CPE) courses to professionals across the nation.

Wiltz started his IT career in the US Navy as a Cryptologic Technician (CTM). Still active professionally in the IT field, he applies his experience to more than just the datacenter. He has spoken at several conferences on the effective implementation and utilization of technology. Additionally, he founded (*and is past president of*) the Mississippi Technology Users Group (MSTUG), which unites IT personnel to promote efficiency in digital solutions.

A speaker at numerous conferences on the effective implementation and utilization of technology, he is also the co-host of the weekly "*Everyday Tech*" show on MS Public Broadcasting (**www.mpbonline.org/everydaytech**). He utilizes his enthusiasm and understanding to inspire and educate, executing diverse methods to provide a "*real-world*" view of the digital age.

In writing Wiltz strives to explains technical matters for the non-technical, bringing that which he has successfully taught in the classroom to the reader. Using analogies and humor, he explains things such as computer security, Internet safety, and computer repair in a way even his mother-in-law can understand.

Married to Kristen, they have two children (*Wiltz III and Izabella)*, and proudly call Flowood, Mississippi home.

Emergency Contact Information

Below is a handy place to record pertinent information in case of a digital emergency. The worst time to try to look up the information is when things are broken.

Internet Service Provider Tech Support: ____________________________

Online Backup Provider Tech Support: _____________________________

☼ ☼ ☼ ☼ ☼ ☼ ☼ ☼

_____ _____ _____ _____ _____ _____ _____ _____

Lights on Cable/DSL//Satellite Modem

Indicate label, color, and blinking or not

☼ ☼ ☼ ☼ ☼ ☼ ☼ ☼

_____ _____ _____ _____ _____ _____ _____ _____

Lights on Wireless Equipment

Indicate label, color, and blinking or not

www.ingramcontent.com/pod-product-compliance
Lightning Source LLC
LaVergne TN
LVHW010934110826
845149LV00013B/2594